To Ralph Copleman (1945-2011) for expanding my interest in sustainability by inspiring me to focus on the scientific story of the universe

....

He dreamed great dreams

Including an interview with Freeman J. Dyson

Star Gazing to Sustainability

Appreciating the Scientific Process

Jonathan Yavelow, Ph.D.

Kendall Hunt
publishing company

Kendall Hunt
publishing company
www.kendallhunt.com
Send all inquiries to:
4050 Westmark Drive
Dubuque, IA 52004-1840

ISBN 978-1-4652-1921-3

Printed in the United States of America
10 9 8 7 6 5 4 3 2

CONTENTS

PREFACE

There is a crisis in both science appreciation and science literacy in students and the general public. This crisis undermines the ability of society to deal rationally with its societal and environmental problems. *Star Gazing to Sustainability* conveys how scientific thinking has uncovered our story of the universe. It is a jargon-free science story celebrating the beauty and power of scientific thought. This book differs from traditional texts in that it includes all of the sciences and focuses on science appreciation rather than going in-depth into any one of the sciences. Historical and scientific information from all of the sciences comes together into the scientific story of the universe and the origin of life. Descriptions of experiments with data from each of the science disciplines are set off in boxes so as not to interrupt the flow of the book. These sections focus on how we know certain scientific facts rather than just what we know. The last chapter presents scientific thinking as central to developing solutions and strategies to the problem of ecological sustainability.

This book is a short introduction to all of the sciences and it is anticipated that students will use information in this book as a start for more in-depth study. There are questions to consider designed to help students start their own independent projects. A bibliography consisting of books, web sites, and videos is provided in astronomy, chemistry, geology, biology, and sustainability. The book ends with an interview with Freeman Dyson thus providing the reader with a window into the mind of one of the most eminent scientists of the twentieth century.

Science appreciation precedes science literacy. There is both joy and insight in science. This book describes the history of the sciences with this in mind. Beyond the knowledge of facts and methods, science literacy requires empowering yourself to think critically. I believe that if people become science literate and are confident in asking, "How do you know . . .?", then we will build a stronger participatory democracy.

When many people hear scientific words (even those describing various fields of science such as astronomy, chemistry, geology, biology, or sustainability), they are already turned off. Often, their experience is that learning science and reading science books are impossible, boring, and exhausting activities.

Star Gazing to Sustainability presents scientific information by using the unique approach of integrating what we know, how we know it, and how it affects our perspective on the world. This approach increases the reader's science appreciation, science literacy, and self-confidence. Science is an iterative process; if data are obtained that disprove a hypothesis, then the hypothesis needs to be modified. This is humbling to the scientist, but also empowering to anyone who practices critical thinking. To encourage critical reading, reflections are incorporated in the book. I suspect these thoughts may also be occurring in the mind of the reader.

The world in which we live can defy our senses. The earth spins at 1,000 miles/hour, yet it doesn't feel as if the earth is spinning at all. Also, the earth is traveling at an average rate of 66,000 miles/hour to complete one revolution around the sun in one year. Again, we don't feel that we are moving, but we are. Scientific insights from astronomy, chemistry, geology, and biology yield a story. It begins with the stars and the origin of the chemical elements, progresses to a supernova explosion and the subsequent formation of our solar system and the earth as a dynamic planet capable of sustaining and evolving life.

What is sustainability and why should we behave sustainably? It is more than recycling and planting trees. It is a mindset linking our understanding of the global ecology of the natural world with good planning for the future. The science-based story of the universe and our origins leaves no doubt that we are intimately part of nature and the environment is part of us. Plants take in carbon dioxide from the air and release oxygen; animals take in oxygen and release carbon dioxide. Our burning of fossil fuels (coal, gas, and oil) sustains our energy-rich life styles and increases the carbon dioxide in the atmosphere. Even though this is hard to believe—we are affecting the global climate.

Why can we believe a scientific fact more than a random guess or a whim? The process of criticism, debate, theoretical understanding, and experimental verification adds to the credibility of scientific facts. Credibility comes not just from the scientific method of observation, hypothesis, experimentation, and conclusion. It is also comes from the community of scientists that verifies the observations and debates the conclusions. After many years, the information continues to accumulate so scientific theory becomes scientific fact. The scientific process is always open to falsification so, at some point in the future, we may generate data that will require us to change our thinking. The history of science shows this to be true. Quoting Albert Einstein:

"One thing I have learned in a long life: that all our science, measured against reality, is primitive and childlike – and yet it is the most precious thing we have."

ACKNOWLEDGMENTS

The idea for this book was inspired by the late Ralph Copleman with whom I had numerous conversations on how to communicate the story of the universe to college students. We taught a course together at Rider University in Fall 2010 titled, *The Environment: A Conflict of Interests*. He always said that in order to bring people together with differing opinions about the relationship of humans to the environment, we need to teach the Universe Story. Ralph then introduced me to Sister Miriam MacGillis at Genesis Farm in Blairstown New Jersey who has been a continuing source of inspiration. Both Ralph and Miriam have fostered in me a deep commitment to Sustainability.

I greatly appreciate Rider University for providing me with a fruitful work environment for more than 30 years as well as supporting this project with Summer Research Fellowships both 2011 and 2012.

This book was written when I was a visiting scholar at the Institute for Advanced Study in Princeton, NJ for the 2011-2012 academic year. My desk was on the first floor of the History and Social Sciences Library overlooking the Institute pond and woods. The library atmosphere was wonderful. I am thankful to one of the Institute's librarians, Marcia Tucker, for her help accessing materials, particularly from the Rare Books Collection. I am profoundly grateful to Freeman Dyson, noted theoretical physicist and mathematician and friend, for many enlightening conversations and inviting me to the Institute.

Drafts of this book have been used in my classes for non-science majors and students have provided reflections on their thoughts. Those reflections are gratefully appreciated. Input from Dr. Bryan Spiegelberg, with whom I taught *From the Big Bang to the Origin of Life*, was also extremely helpful. The editorial work of Melissa Lavenz and Beth Trowbridge at Kendall Hunt is also gratefully appreciated.

Constructive input from the following people is also gratefully acknowledged: Barbara Bluestone, for her many insightful comments, encouragement, and advice; my brother, the late, Mark Yavelow, for his enthusiastic support; Steve Bluestone, for encouragement in the early phase of this project; Tim McGee, for his insights into the history of western thought; Jonathan Millen, for advocating the interdisciplinary science approach and its value to all university non-science majors; Marilyn Gilbert and Ann Marie Hill for their insightful criticism and expert edits; my father-in-law Gene Kreves for testing the strength of my ideas; James Thomashower and Michael Brogan for encouraging me to expand the sustainability chapter; Dr. Mary Leck for her thoughtful comments about botany as well as ideas for reorganizing pieces of the text; David Mannes and my daughter, Ivia Sky Yavelow, for providing me with their perspective of contemporary college students.

Writing even a short book is a significant effort. Throughout all of the ups and downs I gratefully thank the unqualified love and support of my wife, Joy Kreves. Her open and honest critiques have improved every page.

INTRODUCTION

Stories bind people together. They communicate common experiences and values and might bring events of the past alive and into the present. For many of us, there is little better than being totally absorbed in a good story. The story might be written or spoken, and the events true or imagined. The truth of a story often doesn't matter if it builds community spirit or entertains. But in certain instances, the truth of a story does matter. This is particularly important if the truth can be used to predict or plan for the future.

Stories about the universe and our origins have been told for thousands of years and creation stories that all of us know are deeply embedded in our culture. This book is a science story celebrating the universe and our origins. It is my hope that you will read this book slowly. By doing so you can take the information from all of the sciences and reflect upon it. The result will be an enhanced appreciation for both the process of science and the information we have learned from it about the natural world as well as ourselves. Science is about uncovering mysteries. Though there is much debate within the scientific community over the quality of experimental results and what they mean, we know that over time, scientific facts to which we can all agree do emerge. The scientific processes of data-based decision-making and constructive criticism are methods that deserve the utmost respect by our culture.

Imagine that we all add a new story to our lives. This story would be a complement to everyone's emotional, social, religious, and professional stories. It would be the consensus story that has emerged from scientific insights obtained over the last 400 years. This story is a mixture of knowledge and imagination. Many know and believe this story. It begins with the Big Bang and proceeds through the origin of the galaxies, the origin of the elements, and the origin of our sun, earth, and solar system about 4.5 billion years ago. Then, there was the origin of life, evolution of bacteria, viruses, plants and animals, creation of ecosystems and ultimately, the origin of humans and conscious awareness. As you learn more about this story, it is my hope that you will feel more connected to everything around you in the world. I hope the story of the universe and our origins increases your sense of happiness and well being because it provides you with a start to answering the question, " Where do we come from?" I have included brief histories of astronomy, chemistry, geology, and biology for the purpose of giving you a sense of how far we have come in our thinking over the last several hundred years. The accumulation of knowledge in the different sciences has been achieved through independent processes. Although there are gaps and loose ends, most of the information fits beautifully together into this story.

It was 1986. I was in the hall in front of my laboratory in the science building at Rider University. A college student was wearing a Bart Simpson tee shirt that said, "Underachiever and Proud of It." I called the student over to me and said, "That tee shirt is really insulting. I can't believe you are wearing it!" He said, "Oh, Dr. Yavelow, lighten up! It's only a tee shirt." Twenty-five years later, this motto still irks me as the story of underachievement continues. What is an educator to do?

Here is a list that underscores the power of clear thinking, hard work, and accomplishment. This is how I aspire for all of us to be educated.

We listen and we hear.

We read and we understand.

We can talk with anyone.

We can write clearly and persuasively and movingly.

We can solve a wide variety of puzzles and problems.

We respect rigor not so much for its own sake but as a way of seeking truth.

We practice humility, tolerance, and self-criticism.

We understand how to get things done in the world.

We nurture and empower the people around us.

We are able to see connections that allow one to make sense of the world and act within it in creative ways.[1]

What is science? Science is a collection of tools that turned out to be useful.[2] By this I mean science is a method of inquiring about the world. It utilizes the tools of mathematics as well as telescopes, microscopes, and other pieces of scientific equipment to explore nature and test hypotheses. It begins with questions. How far away is a particular star or galaxy? What is the origin of the earth and sun? Where do gas and oil come from? What is an earthquake and what causes a tsunami? Where do plants and animals come from? Where does the air come from? What is the relationship between humans and other animals on the earth? How might degradation of the environment affect life on earth?

After you figure out what question to ask, you think about possible answers and then design experiments and make observations to see if your answer is correct. The act of coming up with your own question and answering it is incredibly empowering. You are trying to determine what is true. When studying how the universe works science has uncovered explanations previously attributed to God. Thus, science has influenced religion, even though the pursuit of science is not religion. So, science is not just about the information that is learned. It is also about the ability to ask the right question and to figure out how to answer it. Galileo (proving that we live in a sun-centered solar system) and Newton (developing the theory of gravity and inventing calculus) made huge contributions to science and were also devout Christians. Science is both the collection of tools and the resultant facts that have helped us to better understand the world.

Here is an experiment for anyone to do that will exemplify the scientific method at work. It begins with you. The question that I am asking you to consider is what is the function of the silks in an ear of corn? My assumption is that by carefully looking at the ear of corn we will find clues to our answer. Take an ear of corn and begin to slowly take off the leaves of the husk one layer at a time. This will require you to pay attention to the task. As you get closer to the corn kernels with the silks above them continue to be careful. Now, with the husk removed, notice where the silks enter the corncob. Each corn silk enters the cob at a kernel. What does this mean? The silk functions as a path down which a pollen grain travels from the air onto the developing cob. There, the pollen fertilizes the ovule (female gamete), creating the seed from which the corn kernel develops. From the myriad of pollen grains blowing in the wind, only one follows each silk strand and is successful in creating a corn kernel!

Do silks in an ear of corn have to have a function? Yes! The thinking in contemporary biology is that without a function the silks would not exist because they would not be selected for by evolution. We can generalize this to everything that we see in the biological world having been selected for over millions of years of evolution. All plants, animals, insects, fish, and other forms of life have evolved from the same mechanism.

Scientific pursuit is a habit of mind that leads us to make up our own minds on a particular subject. When someone tells you something or you hear it on the radio or TV, ask yourself, "How do they know?" In this sense science is self-empowering. To be scientifically literate is to understand and use the methods of science, not just to know the scientific facts.

The pursuit of knowledge in the different areas of science is similar. It is hard work, can be very intense, requires attention to detail, can be disappointing when we are wrong, and can be exhilarating when we are right. For all scientists the discovery is the true reward.

Consider the astrophysicist who identifies spectra in stars that are not present on earth and theorizes that stars contain a new chemical element.

Consider the chemist who invents a new metal alloy that is stronger than steel.

Consider the geologist who discovers a new way to generate natural gas under certain pressure and temperature conditions.

Consider the biologist who realizes the importance of wetland environments to detoxify environmental pollutants.

The practice of science among all of the disciplines is about asking and answering questions. Over the last 400 years this rigorous process has forged a story of the universe and our origins; although it still contains some gaps the story is remarkably coherent.

What do *you* think about the universe, the origin of life and evolution? The origin of life is still one of the gaps in the story. In the course of your lives you certainly have had some ideas on these subjects. These thoughts act as filters through which we all learn and evaluate what we hear, see and believe. As you read please try to be conscious of your biases because this will help you be open to potentially important new ideas. Remember, the Universe Story can be added to all of the other stories of your lives.

I had the opportunity to take a course at Genesis Farm, Blairstown, New Jersey, taught by Sister Miriam MacGillis during the summer of 2010. The focus of the course was the New Cosmology. It was the scientific view of our origins, and its importance for Miriam was not clear to me at first. One of our projects was to create and walk on a time line that spanned from the big bang to the present day. This was a walking meditation. As we proceeded along the time line we listened to a reading about the critical events in the "deep history" (13.7 billion years) of time; for example, the big bang, or the origin of the sun and our solar system, or the evolution of life. After this exercise was finished, we had a few hours to walk around the 225-acre farm and be alone with our thoughts.

> *"In the span of just a few thousand years, teaches Sister Miriam, humans have polluted the waters, altered the global climate and*

driven countless species to extinction. A big part of the problem, she says, is that industrialized Western civilization has believed the wrong story all along – the one in which we can take anything we want from the earth without any bad consequences.

"At Genesis Farm, people come to learn an alternative cosmology – one based on the interconnectedness of all life – and then talk about what can be done to confront an ecological crisis that isn't going away. Sister Miriam calls the hard work ahead "the greatest spiritual quest any generation has ever been asked to take."[3]

Miriam MacGillis was a student and friend of Thomas Berry who believed that there could not be peace among humans without our living in peace with our earth environment. The New Story according to Berry was about having a "comprehensive ethics of reverence for all life."[4] Berry thought we need to find and become connected to a poetic wisdom derived from the story of the universe. Our ability for self-reflection puts humans in a unique position of personal responsibility for the continuation of the evolutionary process. Berry closes his essay on "The New Story" with a passage evoking a confidence in the future despite the tragedies of the present. He writes:

"If the dynamics of the universe from the beginning shaped the course of the heavens, lighted the sun and formed the earth, if this same dynamism brought forth the continents and seas and atmosphere, if it awakened life in the primordial cell and then brought into being the unnumbered variety of living beings, and finally brought us into being and guided us safely through the turbulent centuries, there is reason to believe that this same guiding process is precisely what has awakened in us our present understanding of ourselves and our relation to this stupendous process. Sensitized to such guidance from the very structure and functioning of the universe, we can have confidence in the future that awaits the human venture." [5]

Here is the humanist viewpoint as articulated by Stephen Weldon in the *History of Science and Religion in the Western Tradition*:

"Humanism in the late twentieth century is an anti-supernaturalistic worldview that claims to rely on both the findings and the methods of science. Its ethical system is based on assumptions about individual worth and the ability of human beings to take control of their own lives for the betterment of themselves and those of the rest of humanity. In the humanist worldview, science comes to the aid of humankind,

helping answer fundamental questions about the place of human beings in the world by providing nonreligious answers to age-old religious questions. In the same way that Newton and Darwin became exemplars of the scientific ideal, so, too, have contemporary scientists such as Skinner, Sagan, and the sociobiologist Edward O. Wilson often become exemplars for the humanists precisely because their work provides a cogent reply to the traditional religious explanations of the world. Humanists have praised the utility of science for its ability to provide people with both knowledge and control. With these two possessions, they have repeatedly argued, humankind can take responsibility for its own future." [6]

As the above quote states, humanists believe that science has shown us to have both knowledge and control for ourselves and that we must take responsibility for our future. If the "dynamics of the universe" as articulated by Berry have also prepared us to be capable of the task, then so much the better.

Reflection

The most important thing about science in general for the average person is outlook. The ideas presented came about through imagination, observation, experimentation, and analysis. This is a thought process labeled as the scientific method. From what one can observe, most people don't use this process outside of school unless they become a scientist. What a shame! This is a terrible way to go about living. All it takes is a little knowledge, scientific thought, and open-mindedness. If there is one common thread among the vastly diverse disciplines of science, it is about being open-minded and questioning as much as it is about new discoveries.

This book is organized into chapters on astronomy, chemistry, geology, biology, and sustainability. Scientific information is presented in a historical context to review the evolution of thinking in the variety of disciplines. Specific experiments expose the reader to data on which the science is based.

Think about the difference between what scientists observe and the inferences that are made from these observations. Take time to look at these experiments and you will improve your science literacy. Including all of the sciences in one book is intended to underscore the synergy among these disciplines. The result is a science-based story of the universe and our origins.

Evidence from all of the sciences provides no doubt that humans are part of nature. This realization is a critical first step to dealing more effectively with our environmental problems. My previous students' reflections have been edited and integrated into the text. They provide a personal account of feelings that may also be triggered in your mind. I hope you will realize the power of science and increase your appreciation for data based decision-making. I refer to this as discovering your "inner" scientist. The book ends with students interviewing the renowned theoretical physicist Freeman Dyson. This personalizes the process of science.

I can think of little that is more fulfilling in life than being able to pay forward into the future my joy and knowledge in the hope of building a more conscious, respectful, and sustainable world.

FIGURE 1.1
Johannis Hevelii Machinae Coelestis Pars Prior, Johannes Hevelius (1611-1687). This etching details the beautiful celestial machine from which the astronomer knew the coordinates of the stars and planets he was observing.

CHAPTER ONE Astronomy and the Expanding Universe

The Cosmos is all that is or ever was or ever will be. Our feeblest contemplations of the Cosmos stir us—there is a tingling in the spine, a catch in the voice, a faint sensation, as if a distant memory, of falling from a height. We know we are approaching the greatest of mysteries.[1]

~Carl Sagan

Imagine going outside and gazing up at the stars. Many scientists have dedicated their lives to studying the stars that you are seeing. They have charted their movements and analyzed their properties. Over the years astronomers have mapped the sky with more and more precision. Modern telescopes allow astronomers to see light that our eyes cannot see. These other forms of light (a form of energy) include high-energy cosmic rays, microwaves, and radio waves. What does the light tell us about the story of the universe? Where do these various forms of energy in the sky originate? These questions have been answered and they agree with theories in the field called astrophysics. The theories and observations have given us a very good idea about the history of our universe going back 13.7 billion years!

In 600 BCE, Greek civilization believed the universe was understandable through thought alone. Thales (circa 629-555 BCE) believed that questions about life and the universe could be rationally answered rather than relying upon irrational and subjective myths. In this sense, the Greeks glorified the human mind and are considered the founders of science. Greek cosmology according to Aristotle (384-322 BCE) starts from the structure of the universe bounded by a sphere containing all of the stars (Figure 1.2). According to Aristotle, all time and space exist only within this sphere and there is nothing outside of it. The earth is in the center with the sun, moon, and planets orbiting around it. The boundary of stars is divine, fixed, and eternal. Thus, there is an unchanging celestial sphere. Different from anything on earth is the ether, a sublime element causing light emission from the stars. In the universe's most inner region of the earth and the moon, there exists a world of change evidenced by the changes of worldly weather.

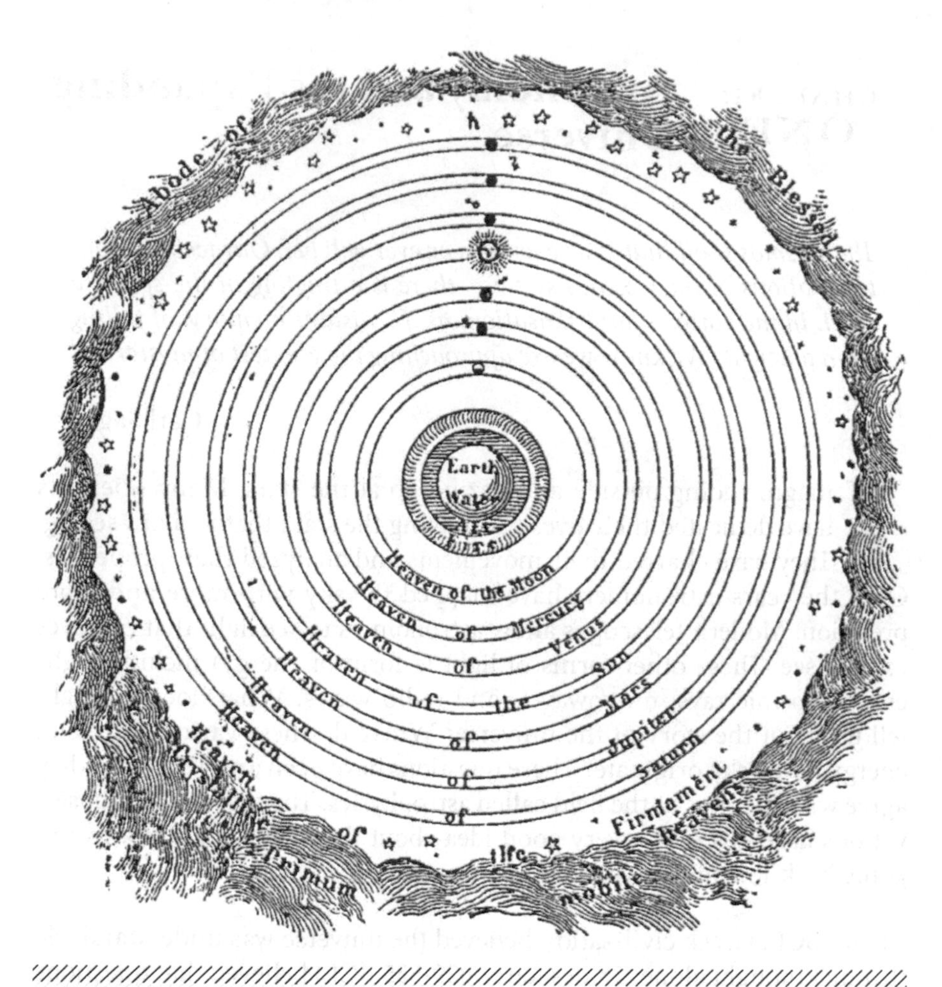

FIGURE 1.2

The Aristotelian view of the universe. The earth is in the center with the moon, planets, and sun revolving around the earth. In the sky are the 'Crystalline Heavens" and "Abode of the Blessed." This was the dominant theory of the universe for about 2000 years. Chinese, Indian, and Egyptian cultures held similar views of the universe.

Now let's think about how observational astronomy was performed before telescopes. First, envision a long rod that you can point toward stars in the sky. This rod is attached to a spherical device with numbers marked upon it so you can precisely determine the angle from vertical to which the rod is pointing. Using this instrument, Hipparchus (circa 194-120 BCE) mapped the position of 1080 stars over thirty years, gave estimates of their brightness, and charted the movements of the planets. This sky chart provided the basis for observational astronomy. In contrast to the movement of the stars, which follow straight lines when moving through the night sky, the planets moved and then reversed direction.

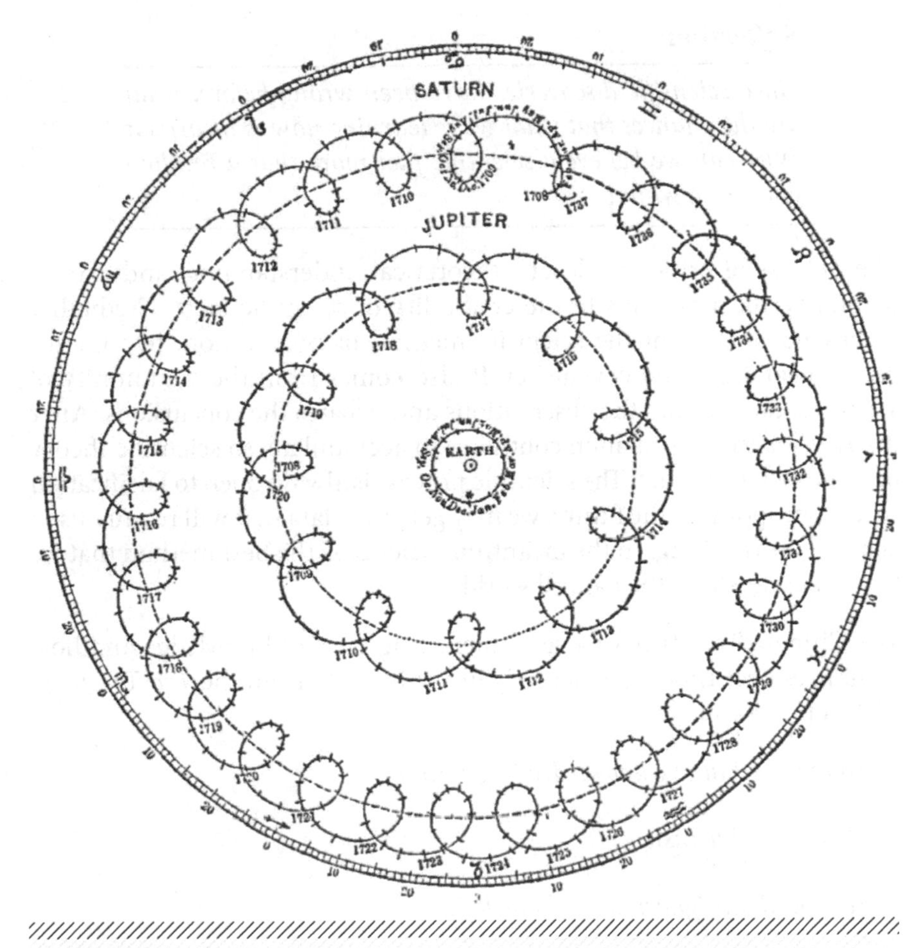

FIGURE 1.3

Epicycles were necessary to account for the movements of Jupiter and Saturn.

Hipparchus introduced the idea of epicycles (Figure 1.3), which are secondary circular motions, in order to account for these observed planetary movements.

Ptolemy's (87 CE-150 CE) mathematical work, *Almagest* (translated from Arabic meaning "The Greatest"), was the book used to predict the position of the sun, moon, planets, and stars. *Almagest* successfully predicted lunar and solar eclipses leading to its acceptance as the truth for fourteen centuries. Imagine how difficult it would be to replace a view of the world/universe that was so deeply engrained in culture. This view of the world was shown by Copernicus (1473-1543), Galileo (1564-1642), and others to be false and was ultimately corrected to a sun-centered (heliocentric) model of the universe.

Reflection

Since scientific discoveries have been wrong before, what are the chances that what we're learning now is incorrect? Why can we believe a scientific fact more that a random guess or a whim?

The process of criticism, debate, theoretical understanding, and experimental verification adds to the credibility of scientific facts. Credibility comes not just from the scientific method of observation, hypothesis, experimentation, and conclusion. It also comes from the community of scientists that verifies the observations and debates the conclusions. After many years the information continues to accumulate, so scientific theory becomes scientific fact. The scientific process is always open to falsification so, at some point in the future, we may generate data that will require us to change our thinking. In the meantime, science is the best method that we have to understand the natural world.

The Chinese Taoists have a very interesting view of knowledge and how it changes over time. I particularly like this quote from the *Tao Te Ching*, Chapter/Verse 71:

To realize that our knowledge is ignorance,

This is a noble insight.

To regard our ignorance as knowledge,

This is mental sickness. [2]

The *Tao Te Ching* is literally translated as "the way, strength/virtue, scripture" and was written in China about 2500 years ago by Lao Tzu. The philosophy of Taoism, based on the *Tao Te Ching*, states that harmony is achieved when opposites are in balance (Yin and Yang). This also includes the balance between the intellect and intuition. According to the *Tao Te Ching*, our effort to objectively learn about the natural world is most complete when balanced with an appreciation of the beauty and mystery of nature.

Humans have come a long way since Rome in the 1600s, when Father Giordano Bruno was tortured and burned at the stake for his beliefs. Poor man! There is a question whether Bruno was killed for his theology instead of his science, but he certainly had some wild ideas for his time.

His observations and thoughts had led him to conclude that there were many inhabited worlds. He stated:

> *In space there are countless constellations, suns, and planets; we see only the suns because they give light; the planets remain invisible, for they are small and dark. There are also numberless earths circling around their suns, no worse and no less than this globe of ours. For no reasonable mind can assume that heavenly bodies that may be far more magnificent than ours would not bear upon them creatures similar or even superior to those upon our human earth."* [3]

We have not yet found life elsewhere in the universe, but Bruno's statement has been accepted by astronomers. Imagine the fear that many scientists of the time, particularly in Italy, must have felt if their thinking was not in support of the church. This is why Copernicus published his work only in the same year as his death and why Kepler's teacher kept teaching that the earth was the center of the universe even though he admitted to Kepler that he actually did not believe it. It takes courage to be a scientist and to follow your own convictions. This remains as true today as it was 400 years ago.

Five scientists who brought forth the revolution in astronomy during the Renaissance were Nicolaus Copernicus (1473-1543), Tycho Brahe (1546-1601), Johannes Kepler (1571-1630), Galileo Galilei (1564-1642), and Isaac Newton (1642-1727). The following is a brief review of what each of these geniuses contributed to science.

Copernicus developed a new quantitative, heliocentric cosmology where the sun is in the center of our solar system (Figure 1.4). This diagram moved the earth (and thus human beings) out of the center and contradicted church teachings. As a result, *De Revolutionibus Orbium Coelestium* was banned by the Catholic Church from 1616 until 1822!

With more precise instrumentation, Brahe made a new map of the sky (see illustration in front of chapter) and updated Hipparchus's map of the heavens. In 1572 he had the good fortune to observe a supernova explosion, a star that was visible on one day and literally exploded on the next! This was the first known documented observation of a supernova since the Chinese observed and documented the supernova of 1054. These supernova explosions were in our Milky Way galaxy. Brahe then concluded that Aristotle's view of an unchanging celestial sphere must be wrong. Labels on the Aristotelian view of the universe (such as "Abode of the Blessed" and "Crystalline Heavens") were challenged. A second blow to Aristotle's

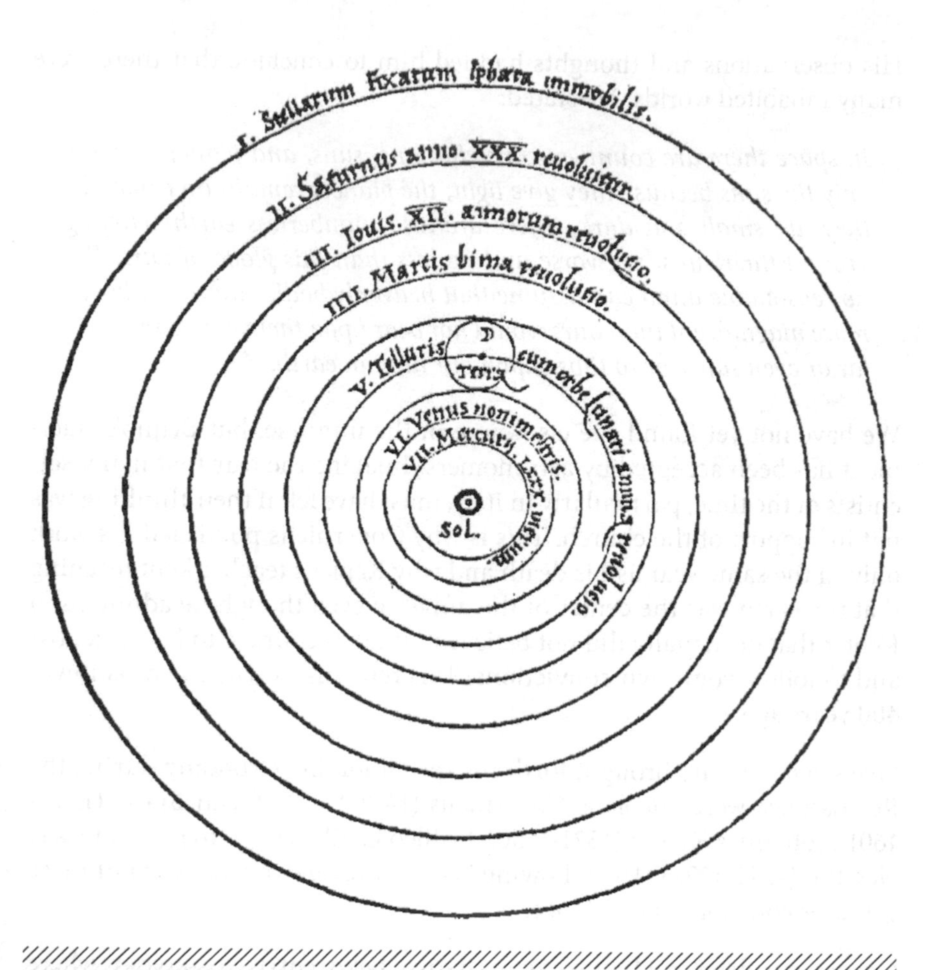

FIGURE 1.4

From *De Revolutionibus Orbium Coelestium*, by Nicolaus Copernicus 1543. In this diagram, the sun is in the center of our solar system and the earth (Terra) has the moon orbiting around it.

view occurred when Brahe observed a comet in 1577. Again the celestial sphere was seen to be changing, not static. In spite of all of his observations, Brahe did not embrace the Copernican view of the solar system. It is difficult to believe that your own data overturn a view of the world that most people believe. It is no wonder that entrenched ideas are hard to forget – even for scientists. Science, in 1577, was already challenging prevailing ideas of what the universe is and where we come from.

Kepler developed the laws of planetary motion using the data of Brahe. He worked at Brahe's observatory, but it was only after Brahe's death that Kepler obtained all of Brahe's meticulous data that charted positions of the stars and planets. He studied Brahe's data on the movement of Mars and reasoned that he had to correct its movement to account for the movement of the earth. Ultimately, after six years of calculations and more than 1000 pages, he succeeded in determining both the orbits of Mars and earth. Kepler proposed that planets move in elliptical rather than circular orbits around the sun. From his calculations, he also noted that planets move faster when they are closer to the sun and slower when they are farther away.

Galileo is best known for the trouble he got into by stating that we live in a solar system where the earth revolves around the sun. At age seventy, in 1633, he was requested to appear before a Holy Inquisition because he taught Copernican doctrine and had written his *Dialogo.* His writing consisted of a discussion among a Copernican, an Aristotelian, and an impartial observer that defended heliocentrism. More importantly, the *Dialogo* stated that the ultimate test of a scientific theory could be found in nature. Galileo believed the world could be dealt with objectively. We call his way of thinking the scientific method! The scientists' view of the universe was challenging the authority of the church. One hundred years after Copernicus, the debates were still raging. Galileo was aware of what had happened to Father Bruno, so in order to save his own head he recanted his own writings.

Galileo's telescope could observe the stars about a hundred times better than the human eye. He saw the mountains of the moon, Jupiter's moons, and many more stars. Later, astronomers recognized some of these stars to be galaxies, each composed of billions of stars.

After Kepler discovered the shape of the orbit of the planets around the sun, Isaac Newton sought to explain it. Why does a planet orbit the sun and not just go off into space? Why does the planet speed up in its orbit when it is closer to the sun? Newton's theory of gravity explained why. The larger the mass of an object, the stronger its gravity and this is why planets orbit around the sun. Newton made no attempt to publish his work and found the pure joy of understanding to be his reward. It was Edmund Halley (of Halley's comet) who paid the cost (fifty pounds sterling) and persuaded Newton to publish, so in 1687 Newton published the *Mathematical Principles of Natural Philosophy.* Newton was a genius and in addition to discovering the theory of gravity, he invented calculus and proved that white light is composed of all the colors of the rainbow. Breaking up light into its component parts enabled astronomers to understand what causes stars to shine (see below).

FIGURE 1.5

A frontispiece by Stefano della Bella for Galileo's *Dialogo*, 1632.

How Do We Know?

CALCULATING DISTANCES TO STARS

How can we find the distance to a typical star in the sky? Here are the mental steps to follow. First, the brightness of a star that we can see with the naked eye is approximately one-trillionth the brightness of the sun. Second, there is a law in physics that the brightness of a star decreases as the distance squared increases. Third, since one million squared is one trillion we can reason that the stars are about one million times further from earth than the sun. The sun is ninety-three million miles away from earth so it is ninety-three million million (or ninety-three trillion) miles to a relatively near star. When we convert this number into light years (which is the distance light travels in one year) we arrive at the number fifteen light years. In other words, it has taken about fifteen years for the light from a near star to reach us. The actual distance of the nearest star is four light-years. This means events on this nearest star that we observe in the sky happened four years ago! When we observe stars in the sky, we are looking back in time.

Let's enter the world of the late 1800s and 1900s when physicists were studying the nature of the atom. As you will see, this information contributed significantly to our understanding of astronomy. Imagine yourself going into the laboratory of a physicist. Glass tubes are filled with purified gases such as hydrogen, helium, nitrogen, and oxygen. These tubes have electric wires enabling the physicist to "excite" the gases with electric sparks. Now envision light shining from the tube of pure hydrogen as it is excited with electricity. Colors of light emanating from the tube can be quantified using a scientific instrument called a spectrometer and each atom has its characteristic pattern of spectral lines. These experiments have been done thousands of times in physics labs around the world and always give the same result. (For more details about the spectral line of hydrogen and the physics underlying the red-shift: http://hyperphysics.phyastr.gsu.edu/hbase/astro/redshf.html#c1)

A startling observation was made when studying the wavelengths of light from the sun through a telescope. Rather than looking through a telescope with your eye, imagine a scientific instrument (a spectrometer) attached to the back of the telescope. The same lines appear as in the laboratory! This is the data that indicates hydrogen is in the sun and is how astronomers can determine the chemical composition of stars. (The actual spectra are more complex because of the extremely high temperature inside stars and the multiple different elements present.)

When further studying the intensity of light from the sun, astronomers were able to determine its temperature. The temperature of the center of the sun is millions of degrees. That is sufficient for nuclear fusion of hydrogen into helium. In other words, energy from the sun is the equivalent in energy of millions of continuous hydrogen bombs exploding. Our sun is like other stars and this is why stars stay so hot and emit light for billions of years.

The hydrogen photograph (Figure 1.6) of the sun was obtained by using a filter that only allowed the light emanating from hydrogen to get through. Notice how dynamic the hydrogen is inside the sun. The swirling pattern results from the intense nuclear reactions occurring continuously. When astronomers looked at the light from far away galaxies, they noticed something different. They saw the characteristic spectral lines of hydrogen but

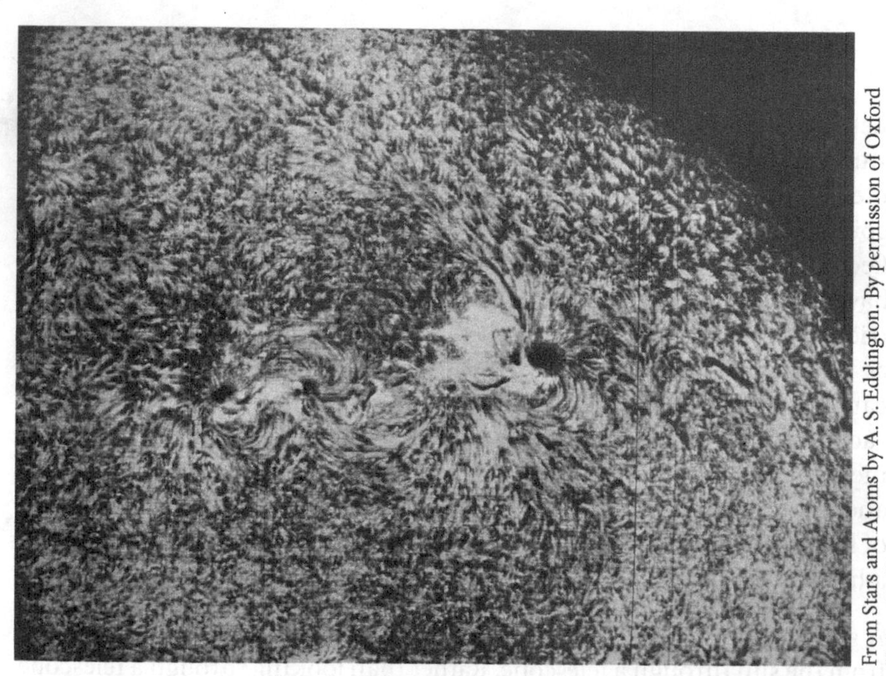

FIGURE 1.6
Hydrogen photograph of the sun.

they were displaced to the right on their spectra (see box below). This red-shift indicated the object was moving away from the observer.

There is a well-known story that when Einstein looked at his equations for relativity, he saw that they suggested that the universe was unstable, but he was not emotionally prepared to believe it. He could have predicted that we live in an unstable universe, but changed his equations with a number (the cosmological constant) that made the equation show the universe as *not* expanding. Such is the power of culture even upon the brightest scientific minds! (Recall that 300 years earlier, Brahe also had difficulty appreciating that his data supported a heliocentric (sun-centered) cosmology.) About twenty years later Einstein saw the results of the red-shifts of the light from galaxies and realized he made a mistake. It was Edwin Hubble (after whom the Hubble Space Telescope is named) who observed that the universe is expanding. He observed that the farther away a galaxy is, the faster it is moving away from us. This is one basis for determining the age of the universe as 13.7 billion years.

With all of this information, how can we get an intuitive feel for the size of the universe? Let's start with imagining our solar system using a metaphor that was created by Martin Rees (Astronomer Royal at Cambridge University):

> *Suppose our star, the sun, were modeled by an orange. The earth would then be a millimeter-sized [approximately 1/32 of an inch] grain twenty meters [yards] away, orbiting around it. Depicted to the same scale, the nearest stars would be 10,000 kilometers [6000 miles] away: that's how thinly spread the matter is in a galaxy like ours. But galaxies are, of course, especially high concentrations of stars.* [4]

A very minor portion of the universe is thought to consist of matter that we can see. Here we can use the metaphor of a Christmas tree with lights. The lights represent the stars and galaxies in the universe and although it cannot be seen at night, the tree certainly has a very strong influence on the arrangement of lights! Most of the matter emits no light, or waves of any kind. It is therefore very hard to detect. How do astronomers know that it exists if it can't be seen? Careful analysis of the movement of galaxies and stars indicate that there is a strong gravitational pull being exerted upon everything that we see. This is referred to as dark matter and, as yet, we know very little about it. As much as we know about the universe, there is much more yet to learn.

How Do We Know?

DATA FOR THE EXPANDING UNIVERSE

The Hubble Diagram is the basis for believing that we live in an expanding universe. Let us take a glimpse at the data from astronomers. Here is a description of the graph and the data on which it is based. The goal here is to simply translate the graphs into sentences that we can understand.

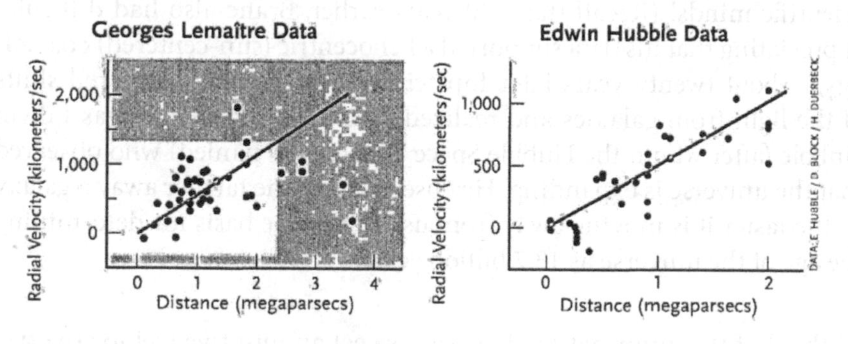

FIGURE 1.7

Diagrams demonstrating that we live in an expanding universe.

Each dot represents a galaxy whose position is determined by the distance to the galaxy (on the horizontal x-axis) and the velocity that the galaxy is moving away from us (on the vertical y-axis). All of the spectra from the galaxies, except for a few that are close to us, are red-shifted. These graphs demonstrate that we live in an expanding universe because most of the galaxies are moving away from us. That statement alone suggests that there must have been a beginning and gives us the impression that we are the center of the universe (for explanation, see below).

There is a region of the sky called the Virgo Cluster of galaxies and it is in this region that our own Milky Way galaxy resides (Figure 1.8). Imagine each smudge of light in the sky consisting of literally billions of stars. Each one of these galaxies is represented as a dot on the graphs above. On the horizontal axis are megaparsecs and on the vertical axis are radial velocities. One parsec is about 3.26 light years (about 19 trillion miles) and one million parsecs is a megaparsec.

How can we determine these huge distances? At this point let's just assume that these distances are correct and feel the awesomeness of the magnitude of these numbers. Now, notice on the vertical axis that the numbers go up to 2000 kilometers/second (1200 miles/second) on the graph on the left and 1000 kilometers/second (600 miles/second) on the graph on the right. In addition to just reading these numbers let's hear what the numbers say: the galaxies in the Virgo Cluster are moving away from us at a rate of hundreds of miles/second. If the particular galaxy is moving at 500 kilometers/second (300 miles/second) then this is equal to about one million miles/hour, yet it just looks like a smudge of light in the sky and appears not to be moving at all! This is because it is so far away. The smudge of light (or galaxy) moving at a speed of one million miles/hour is nineteen million trillion miles away!

FIGURE 1.8
Virgo Cluster of Galaxies.

Let's do some calculations that are a little closer to home. How fast does the earth spin on its axis? The earth is about 24,000 miles in circumference at the equator and there are twenty-four hours in a day, so the earth spins at 1000 miles/hour. It doesn't feel as if the earth is spinning at all, but we know that it is. Similarly, the earth is moving at a rate of 66,000 miles/hour in order to get around the sun in one year. Again, we can't feel that we are moving so rapidly. These simple calculations inform us that the world in which we live can defy our senses. We have come a long way since 600 BCE when the Greeks introduced the idea that the universe was understandable through thought alone!

Consider yourself looking into the night sky. When you see two stars of equal brightness, it is possible that both stars are an equal distance away from earth. Another possibility is that one star is much closer than the other, but because it is intrinsically less bright it appears to have the same brightness as the more distant star. Astronomers have succeeded in distinguishing apparent luminosity (what our eyes see) from intrinsic luminosity (the actual brightness of a star or galaxy). This was deciphered in the early 1900s by Harriett Leavitt and others at the Harvard University Observatory. Night after night Leavitt took pictures of the same region of the night sky and observed that some stars got brighter or fainter at regular intervals (ranging from fifteen to fifty days). Leavitt then made the discovery that the time period between maximum brightness was related to the intrinsic luminosity of the star or galaxy. Other astronomers then independently determined (using geometry) the distance to a particular variable star. Now, Leavitt was able to determine both the distance and luminosity of a star. Using the law that the brightness of a star decreases with the inverse square of distance, she was able to calculate the distance to farther and farther stars or galaxies. All of this was necessary to make the x-axis on the Hubble diagram (Figure 1.7).

How do we determine the velocity of galaxies moving away from us? In order to obtain the data in Figure 1.9, astronomers put two different instruments on the back of the telescope: a spectrometer that breaks up the starlight into its components, and photographic film so pictures of the galaxies could be taken. Now, notice the right-facing arrow under the K and H lines in the velocity column beginning with NGC4473. The space immediately to the left of the horizontal smear is the spectral line from calcium. From studies on earth using pure calcium, we know where this line should be. Yet, when the starlight is analyzed, the spectral line is always displaced to the right. The H and K lines from calcium are displaced to the right and the extent of the displacement is indicative of the speed at which the galaxy is moving away from us. These speeds are written under each panel. On the right of the diagram is the picture of each galaxy and the distance in light years that each galaxy is away from us. In order to get these photographs the shutter on the camera might have been open for an hour. It would be necessary for the shutter speed to be open for the same times in order to compare the luminosity of each picture.

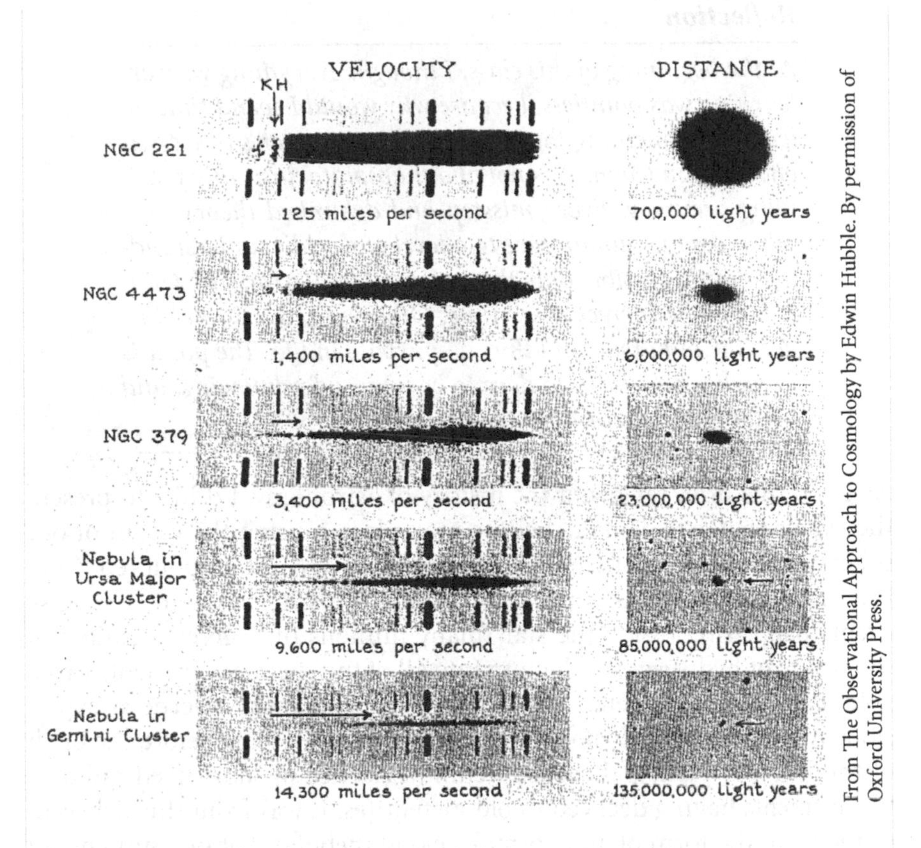

From The Observational Approach to Cosmology by Edwin Hubble. By permission of Oxford University Press.

FIGURE 1.9
Hubble's data showing the velocity and distance measurements of several galaxies.

Lastly, when you look up in the sky at any group of galaxies you can determine both the distance to those galaxies and the speed at which they are moving away from us. If you plot the velocity at which the galaxies are moving away on the y-axis and the distance to those galaxies on the x-axis you **always** get the Hubble diagram (Figure 1.7).

Since most of the galaxies are moving away from us does that make our earth special? It suggests we are the center of the movement. Alas, we think not. We believe the universe is like a balloon with dots on its surface. Each of the dots on a balloon is moving away from every other dot as we blow up the balloon and the farther away each dot is from a particular dot, the more rapidly it appears to be moving away from the dot we are "on." From each vantage point it appears that the universe is expanding away.

Reflection

At the beginning of this class, I thought everything we were learning was pointless, because why would I spend time on learning something that may change in a week's time? Science, I soon learned, is not about absolute, instantly gratifying accuracy. Every misstep and debunked theory only advances the human race to a level of worldly understanding that is possible only through trial and error. The point is not to be correct every time. Otherwise, there would be nothing to pursue and thus no purpose in life. The point is to make that effort, to dare to be hungry for knowing, and better ourselves through understanding.

As I end this brief review of the history of astronomy, I'd like to present the events that astronomers believe took place causing the origin of our solar system. Between 4.5 and 5.0 billion years ago a star about ten times the size of our sun was coming to the end of its life. Extremely intense events happening inside the star (many millions of degrees) resulted in the fusion of hydrogen and helium into all of the chemical elements (iron, gold, carbon, oxygen, etc). This ended with a gigantic supernova explosion. We know the explosion happened only a short time before the birth of the sun and planets because we find evidence of short-lived radioactive elements having decayed in old meteorites. It was from this debris of elements in the form of an enormous cloud (nebula) that our sun and all of the planets formed. The sun (containing 99.9% of all of the visible matter) formed about 4.5 billion years ago when a central blob from the pre-solar nebula condensed to form an object with sufficient mass and gravity to ignite. All of the planets formed from the relatively small amount of leftover material, and the heavier elements that coalesced into dust and particles formed the inner rocky planets. These aggregations had too little matter to ignite like the sun. Thus, the planets were formed. Another 100 billion stars like our sun are in our galaxy alone. (Giordano Bruno was right about this in 1600!)

Below is a picture of a supernova explosion from which astronomers believe new stars are being born. Our sun and the planets were formed from an event like this that most likely happened about five billion years ago.

FIGURE 1.10
Birth of new stars after supernova explosion.

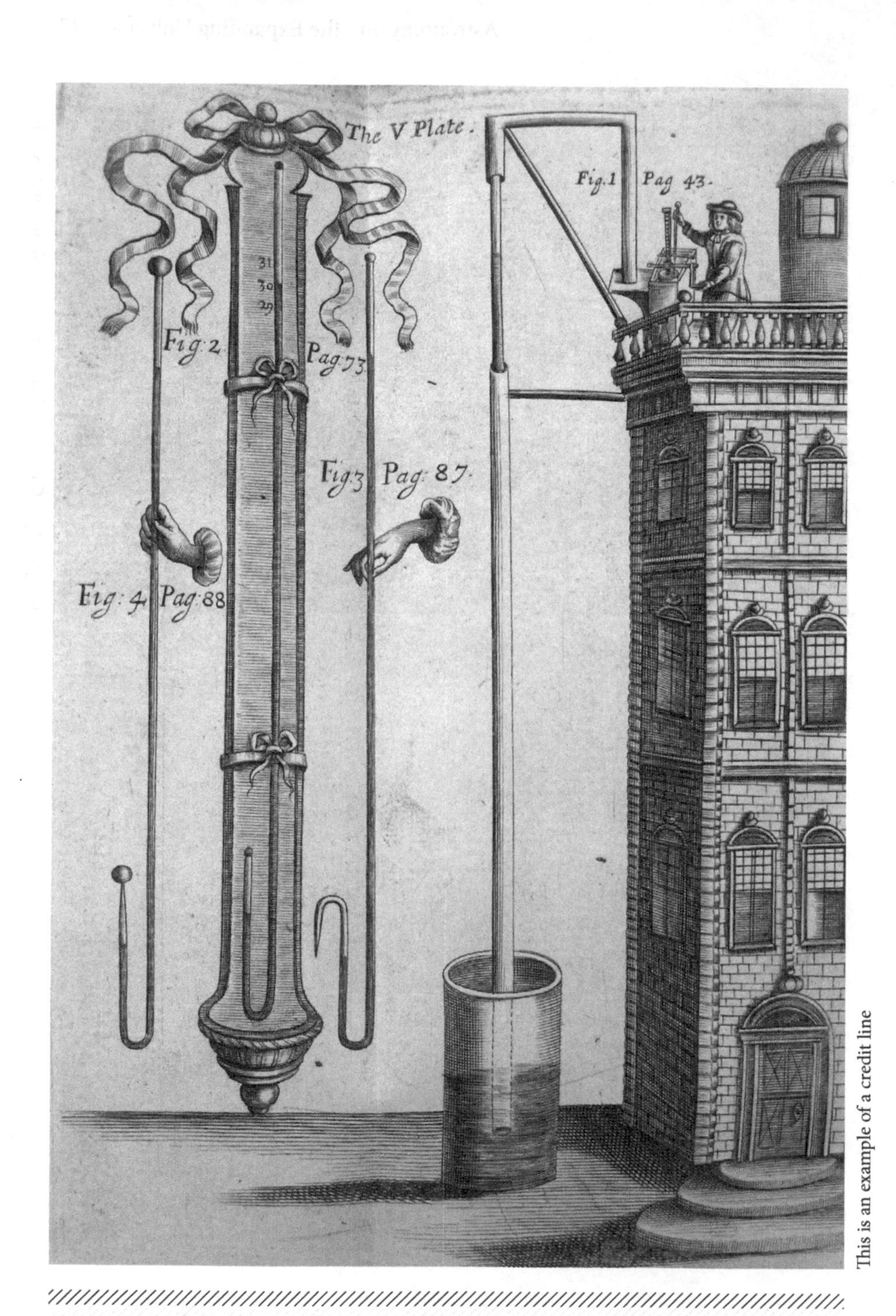

This is an example of a credit line

FIGURE 2.1

Boyle, Robert. A Continuation of New Experiments, Physico Mechanical, Touching the Spring and Weight of the Air, and Their Effects . . . Oxford, Printed by H. Hall, for R. Davis, 1669.

CHAPTER TWO Chemistry and the Origin of the Elements

"It is the stars, The stars above us, govern our conditions" (King Lear, Act IV, Scene 3) but perhaps

"The fault, dear Brutus, is not in our stars, But in ourselves," (Julius Caesar, Act I, Scene 2)[1]

[Do] the stars govern the fate of mankind?[2]

Thousands of years ago, the practice of chemistry was totally different from that of astronomy. Chemists worked in a wet, smelly, smoky environment as compared with the silence of the night in the observatory looking at the stars. Also, chemical knowledge was gained through observation and direct experimentation while astronomy was more about reconciling theory and observation. The chemist might ask, "What will happen if I heat this sample as hot as I possibly can?" whereas the astronomer can simply observe his subjects in the night sky. For chemists, extreme care must be taken so experiments don't explode or release a gas that may sicken the chemist.

In early Sumerian (3000 BCE) and Greek (2500 BCE) civilizations people may have asked themselves the same questions we ask ourselves today. "Where can I find soap and perfume?" "What is the best way to prepare my food?" " What is the best way to purify water so I don't get sick?" "How can I look more beautiful? "Chemistry contributed answers to all of these questions with the creation of soap and perfume, the invention of the process for making cheese, distillation of water, and the smelting of metals for jewelry.

Imagine the beginning of the Iron Age (1500 BCE) when higher temperature kilns for the smelting of iron were developed. The cast iron plow could be attached to an animal for help in plowing the fields so that one animal could do the work of many humans. People freed from working in the fields had time for the development of music, art, science, and culture.

With the development of ovens and crucibles came the desire of some early chemists (alchemists) to transform lead into gold. Many alchemists worked tirelessly in this pursuit and were called "laborers in the fire." Others were less motivated by the material world and wanted to create the magic elixir for eternal life. These experiments often involved heating a reddish ore called cinnabar that resulted in liquid mercury and the release of a smelly sulfur dioxide gas. This discovery coupled with its mysticism prompted the theory that metals were made from sulfur and mercury (subsequently shown to be incorrect). The quest for making the elixir is seen by many today as a fool's errand; however, these laborers in the fire brought forward empiricism and the importance of careful experimentation.

Several cultures (among them Sumerian, Chinese, Indian, Egyptian and Greek) developed a similar framework for chemistry consisting of four qualities and four elements. Fire, earth, water, and air are the four elements and hot, dry, cold, and wet are the four qualities. We can think that earth, water, and air represent solid, liquid, and gas states and fire represents the energy bringing changes among these states.

Are earth, air, and water the basic elements or is nature built from smaller components? This question was first asked by the Greeks. Leucippus (490-? BCE) considered dividing matter into smaller and smaller units. He believed that there was a limit beyond which matter could not be made any smaller. His student, Democritus (460-370 BCE), named these elements *atomos* meaning "unbreakable" in Greek. This was the first expression about the existence of atoms. Several hundred years later in Rome, Lucretius (96-55 BCE) wrote a long poem *De Rerum Natura (On the Nature of Things)* about atomism in 56 BCE. The poem survived for about 1400 years and it was among the first items to be printed using the printing press in 1467. This attests to our long fascination with understanding the basic elements of nature.

The power of chemistry to change society was demonstrated in Europe by Roger Bacon (1220-1292) who made an artificial explosion. He mixed saltpeter, charcoal, and sulfur to make gunpowder. The first explosion was most likely accidental; it probably happened because the ingredients were dry and in the correct proportions. At the same time in China both fireworks and gunpowder were developed from the same three ingredients and as the recipes for gunpowder became more reproducible, chemistry began its legacy of doing both good and harm. Many years later during the Revolutionary War in America, we were fortunate to have the French on our side as they shipped us very high quality gunpowder. This often-overlooked fact was a major factor in our winning the war against Britain.

We all know that science proceeds via experiments but it is often not easy to interpret them. For example, Johannes Baptista von Helmont

(1579-1644) did the following experiment in Belgium in 1615. He took a small, potted willow tree and over a period of five years gave it nothing but water. The weight of the soil changed negligibly over the five years but the tree increased from four to 169 pounds. He concluded that water was all that was needed for his tree to increase in weight. For this reason, he challenged Aristotle's belief in the four elements (earth, air, fire, and water) and stated his experiment proved that only water was necessary for the growth of the tree.

Von Helmont also believed that air was important only for fire. In the early 1600s air was a complete mystery. Many people thought that breathing involved taking in a spiritual aspect from the air that enabled humans to live. It took more than 150 years to discover air was a mixture of gases containing oxygen, nitrogen, and carbon dioxide. Now we believe carbon dioxide in air is the source of the carbon responsible for the increased mass of plants due to the process of photosynthesis.

Robert Boyle (1627-1691) had inventiveness and creativity to make him a successful experimentalist. In 1657, he modified the air pump previously invented by Otto Von Guericke and further investigated the properties of air. The air pump was able to create a partial vacuum in a closed vessel. One of his key modifications was a receptacle enabling subsequent chemists such as Joseph Priestley to test the presence or absence of air on the flame of candles and on the survival of mice. Thus, it was again the development of tools that fostered knowledge about gases. In the illustration beginning this chapter (Figure 2.1), Boyle was on top of the building testing the relationship between pressure and volume. The J-shaped tubes with one open end were used to measure the volume of air trapped in the bottom of the tube. As more mercury was poured into the open end, thus increasing the pressure, the volume of trapped air decreased.

Boyle developed a mathematical relationship (Boyle's Law) among pressure, volume, and temperature and coupled his experimental approach to chemistry with theory. His work represented a coming together of chemistry, mathematics, and physics with practical as well as theoretical consequences to his work. Boyle's Law has been useful to both scientists and engineers in the development and implementation of ideas as varied as municipal water and sewage systems and the internal combustion engine. Boyle was also aware of Lucretius' poem and interpreted his results in the context of atoms. He reasoned that as the pressure increased, the space between atoms of the compressed air decreased.

The power of a vacuum developed by Guericke made quite an impression upon King Ferdinand III. In 1654 in Magdeburg, Germany (Figure 2.2) a famous demonstration was performed on his behalf. Two halves of a large sphere were connected by a vacuum and then two teams of horses tried to pull the sphere apart. They failed. Using Boyle's law we can calculate that if the sphere was two feet in diameter then the air pressure on the sphere would be ten tons! No wonder the horses could not pull the halves of the sphere apart.

FIGURE 2.2

Guericke's demonstration of the power of a vacuum. Otto von Guericke, Experimenta Nova (Ut Vocantur) Magdeburgica De Vacuo Spatio, 1672.

Now fast forward to 1774 and the laboratory of Joseph Priestley (1733-1804) in England. Priestly demonstrated a gaseous and chemical relationship between plants and animals and also plants and combustion. Consider the following series of experiments. A sprig of mint could live inside his apparatus closed to the outside air for months, but a candle or a mouse inside the apparatus would last for only a short time. Why was the plant able to survive for so long compared to the flame or the mouse? Furthermore, a mouse placed into the jar alongside a plant lived happily for ten minutes, but a mouse put inside the jar immediately after another had just expired would convulse in a matter of seconds! From these observations, Priestley concluded that the plant disabled a factor that extinguished the flame or disabled a factor that killed the mouse. He appreciated the connection between plants and animals in the biosphere, but he didn't realize that the plant was giving off something positive (oxygen), showing that it is not easy to break out of the prevailing paradigm of thinking unless one is both ingenious and daring.

It was from the careful experiments of Antoine Lavoisier (1743-1794) that the new era of chemistry was born. A wealthy man, he was able to purchase high quality balances, furnaces, and laboratory glassware with which to perform his experiments. For example, chemists routinely observed that after boiling water in a closed system of glassware small brown specks always appeared. From this observation, they concluded that boiling water caused "earth" to appear. To approach this problem, Lavoisier weighed both the glassware and the water at the beginning of the experiment and proceeded to boil water for several days in a closed vessel. At the end of the experiment he observed the brown specks of "earth." However, upon weighing the water, glassware, and the specks of earth, he concluded that the weight of the specks was exactly the same as the weight lost by the glassware. He concluded that the experimental conditions removed the brown specks from the glass. Lavoisier's careful experimentation thus disproved that earth comes from boiling water. Meticulous attention to detail and note taking contributed to his success.

In his classic experiment proving that oxygen is involved in combustion, Lavoisier heated mercury in a closed apparatus of glassware for several weeks (Figure 2.3, left diagram). He observed that the air decreased in volume by 20% while a red scale like rust simultaneously appeared inside the flask. He correctly interpreted that 20% of air is composed of oxygen, which decreased because it reacted with the mercury. Most of the rest of air, which is nitrogen, was not reactive in this experiment. Based on this observation, Lavoisier reasoned that air is composed of several different gases. Lastly, he heated the red scale to a very high temperature using a magnifying glass to focus light onto it (Figure 2.3, right diagram). It turned the red scale back into liquid mercury and oxygen was released. Lavoisier had demonstrated that gaseous oxygen reacted with the mercury in a reversible manner and interpreted his experiments in light of a new theory of combustion. According to Lavoisier:

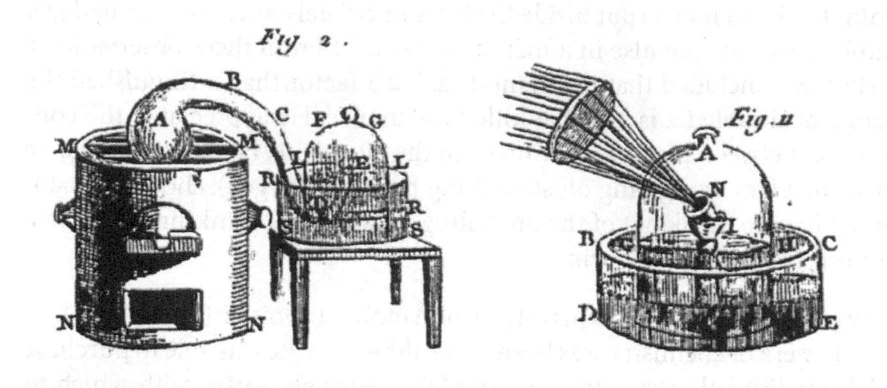

FIGURE 2.3
Lavoisier's glassware used in his discovery of oxygen and combustion.

> *I do not expect that my ideas will be adopted all at once; the human mind adjusts itself to a certain point of view, and those who have regarded nature from one angle, during a portion of their life, can adopt new ideas only with difficulty.* [3]

Prior to Lavoisier, in the seventeenth and eighteenth centuries the prevailing theory of combustion was called phlogiston theory. Phlogiston was theorized to be a substance that was released upon burning. Priestley and other chemists knew that metals gained weight when they were exposed to fire. This data should have refuted phlogiston theory because one should have predicted that the substance would lose weight upon burning. But prevailing theory is a powerful force in the mind of anyone -- including scientists. Lavoisier had courage. His experiments proved that an element in air, oxygen, reacted reversibly with mercury and he had the courage to say that the phlogiston theory was wrong. Now we had a theory of combustion and the element oxygen was established. Chemistry had become an exact science. Later, during the French Revolution, the wealthy Lavoisier found himself on the wrong side from the revolutionaries and he was beheaded in 1794, just a few years after his *Traite Elementaire de Chimie* was published.

Reflection

From the farthest reaches of what can be seen with a telescope to the tiniest molecules that are examined in a chemistry lab, it may seem that most scientific disciplines are distinct and unrelated entities. However, a cross-section of the history of science reveals that the great discoveries were surrounded by great contention. It was those scientists who questioned the status quo that came forth with new innovations–which they were then often forced to defend. If there is one common thread among the vastly diverse disciplines of science, it is about being open-minded and questioning as much as it is about new discoveries.

Finally, chemistry's new conceptual framework based on the atom replaced the four elements of earth, air, water, and fire that had been first proposed nearly 2000 years earlier by Leucippus and Democritus. Any substance that could be broken down was, by definition, not an element. Consider the experiment of Joseph Louis Proust (1754-1826) and see how it is consistent with the concept of matter being made up distinct units called atoms. He took copper carbonate and broke it down into three elements: copper, carbon, and oxygen. The ratios were always five parts copper, four parts oxygen, and one part carbon. If copper was in a shortage, then only the proportionate amount of carbon and oxygen combined with it. This became known as the *law of definite proportions.*

At about the same time, Robert Brown (1773-1858) was sitting at his microscope in 1827 and noticed that pollen grains suspended in water moved without stopping. He thought that perhaps the pollen grains had a"spark of life" in them. To test this thought, he observed objects that had no question of life in them (like glass and coal) and they also moved. This process was subsequently named Brownian motion. James Maxwell (1831-1879) developed mathematics that described atoms and molecules bouncing off each other and finally in 1908 Albert Einstein (1879-1955) applied Maxwell's theory and came up with exactly the mathematical description to describe Brownian motion. He calculated the size of atoms to be one hundred-millionth of a centimeter, 10^{-8} cm, or 250 million atoms lined up side by side would equal one inch. That's small! So atoms are always in motion and that accounted for the pollen grains moving under the microscope.

The experiment of Ernst Rutherford (1871-1937) confirmed not only the existence of atoms but also their structure. His experimental apparatus utilized a radioactive source of uranium which emitted alpha particles (a helium nucleus consisting of two protons and two neutrons). The uranium was surrounded by a lead shield with a slit and it was through that opening that the alpha particles were directed to a target of gold foil. Rutherford anticipated that the nuclei of the gold atoms would deflect the alpha particles, which could then be seen by a detector surrounding the gold foil. In only one out of 20,000 events the alpha particle was deflected back in the direction of the source. This means that 19,999 times out of 20,000 there was one type of result and 1/20,000 there was another. Most of us would discount this 1/20,000 event, but Rutherford calculated that the size of the nucleus of the gold atom was very small and this would be in the range of expectation. Consider the metaphor of a pea inside a baseball stadium to give a sense of the relative sizes of the nucleus to the atom as a whole. In agreement with Einstein, he estimated the diameter of the atom of gold to be 10^{-8} cm and the size of the nucleus to be 10^{-13} cm (1/10 trillionth of a cm).

> *"Yet people still speak of the atomic theory, because that is what it is – an intellectual map of large aspects of science that can be neatly explained by the existence of atoms. A theory, remember, is not a "guess," and no sane and qualified scientist can doubt that atoms exist. This aspect of the proof that atoms exist is also true of other well-established scientific theories. The fact that they are theories does not make them uncertain, even when various fine details are still under dispute. This is particularly true of the theory of evolution, which is under constant attack from people who are either ignorant of science or, worse, who allow their superstitions to overcome what knowledge they might have."* [4]
>
> ~Isaac Asimov

By the turn of the twentieth century, chemists were studying the composition of cells and the field of biochemistry was born. Molecules necessary for life, such as glucose and amino acids, were identified in plants and animals. Furthermore, laboratory simulations mimicking the atmosphere of the early earth (thought to be composed of carbon dioxide, methane, ammonia, and water vapor) resulted in the synthesis of amino acids that are found in all living cells. This is the famous Miller and Urey experiment. Thus, the basic simple chemicals that make up life could have been formed by the chemistry of the primitive earth. The walls were breaking down between chemistry and biology.

More evidence concerning the power of empiricism comes from the discovery of nuclear energy. Antoine Henri Becquerel (1852-1908) discovered radioactivity in uranium. This was followed up by the work of his graduate student, Marie Curie (1867-1934). She worked in Paris with her husband Pierre Curie during the turn of the twentieth century and was looking for the source of heat from a uranium ore called pitchblend. As she continued to purify the source of the heat she succeeded in identifying a new chemical element – radium. Much to her surprise, the heat energy was emanating from inside the atoms. Thus, a new and unexpected natural process was discovered, termed nuclear energy. A few years later, Rutherford discovered the nucleus. After winning two Nobel Prizes, one each in chemistry and physics, Marie Curie died of leukemia. At the time it was not known that radiation emitted from nuclei could cause cancer.

How Do We Know?

DATA FOR THE ORIGIN OF THE ELEMENTS

Between 1900 and 1950, a great deal was learned about nuclear reactions. It was out of this knowledge that nuclear weapons were developed and deployed in 1945. By splitting the atom, the immense power of nuclear energy was released. The idea that one element could be converted into another both by nuclear fission (the breakdown of uranium into lighter elements) and by nuclear fusion (conversion of hydrogen into helium) was the basis for the power of nuclear energy.

Might nuclear fusion occur inside stars causing the stars to shine? As a by-product might these reactions result in the synthesis of the chemical elements? Techniques were developed to monitor these nuclear reactions and when similar techniques were applied to looking at the light from stars some recognizable patterns emerged. Recall that when astronomers look at stars they don't look into telescopes with their naked eye. At the back of the telescope is an instrument named a spectrometer. Inside the spectrometer is a prism (or a grating) that breaks up the light into a series of colors that appear as lines. Think of a prism with "white" light entering at one side and all of the colors of the rainbow emerging from the other side. The colors (or lines) are a "signature" of the particular elements. In this way astronomers assigned certain elements to be present in the stars (for more information see http://www.aip.org/history/cosmology/tools/tools-spectroscopy.htm). Imagine the excitement that you would feel upon discovering that the chemical elements that make up yourself are also present in the most distant stars in the sky!

These complex patterns can be simplified by comparison to spectra from known elements. The sciences of physics and chemistry were coming together.

Reflection

We are all made of stardust, which kind of raises my self-esteem, even though that doesn't make sense. I don't know. I feel prettier.

To summarize the history of chemistry: first, chemistry has provided knowledge for tools that have contributed to the development of human civilization; second, these tools had both positive and negative influences on society; third, knowledge of the chemical elements and molecules are key to the scientifically based story of the universe and our origins. This is because it is from the universe that atoms are made and molecules (composed of atoms) are the materials from which living cells are made.

Science is amoral – neither good nor bad – and is a method for the pursuit of knowledge. Marie Curie and other physicists and chemists responsible for the technology that grew out of their scientific work were following their innate curiosity and intellectual passion. It is technology and society that determines how insights from the sciences are applied. This idea underscores the importance of scientific literacy among our citizens as well as the ideals of a well-educated citizenry with political engagement. Data-based decision-making coupled with open and respectful dialogue and personal responsibility are keys to making informed unbiased decisions in our society.

To end the Chemistry chapter, consider the following story about a carbon atom, retold from Primo Levi[5]:

The story begins with the mining of limestone (calcium carbonate) from the side of a cliff. It is heated it to 900 degrees centigrade, in a lime kiln forming CaO (lime) and carbon dioxide. The carbon dioxide is released into the atmosphere because it has been converted from a solid into a gas. After about twelve years, the carbon dioxide lands on a grape vine leaf at precisely the same moment as a ray of sunshine hits the leaf. Now the carbon is no longer flying in the air; it has become part of a plant and is in a solid form again. It only took one millionth of a second! Compared to oxygen and nitrogen (which represent 21% and 78% of air, respectively) the carbon dioxide only represents 0.03% of the air.

The carbon atom in the plant becomes linked to five other carbon atoms and forms the sugar, glucose. Now the carbon atom flows as sap to nourish the plant and enters the almost ripe bunch of grapes. Imagine that the grapes are harvested and made into wine and the carbon atom becomes part of the alcohol in wine. After the wine is bottled and sold, it is consumed by an individual and the carbon atom becomes stored as sugar in his liver for about one week. Then, while the individual is jogging, the sugar becomes converted back into carbon dioxide and through his breathing is released into the air again. The wind carries the carbon dioxide around the globe and after another twelve years it again lands upon a leaf at exactly the same moment as a ray of sunshine. Again, it is transformed into the plant, but this time it is a cedar tree and the carbon atom becomes part of the bark of the tree. Now let's assume that a caterpillar digs a tunnel between the tree and its bark and swallows our carbon atom after it was in the tree for about twenty years. It forms a pupa and later transforms into part of the eye of a grey moth. After a few months, the moth dies. Now the dead moth becomes buried in dead leaves and becomes converted back into carbon dioxide by the microbes in the soil.

Our carbon atom started in a limestone cliff as part of calcium carbonate and after it was burned in a kiln, it was released into the atmosphere as carbon dioxide. It was incorporated into a grape vine and made into grapes. The carbon atom was then converted into alcohol when it was made into wine and was consumed by a person. After spending one week as glucose, the carbon atom was released into the air again by metabolism. After about twelve more years, the carbon atom became incorporated into a cedar tree where it was part of the bark (cellulose) for twenty years. Then a caterpillar took some of the bark and our carbon atom became part of a moth. The moth died and then the carbon atom was recycled into the air again by soil microbes.

The number of carbon atoms in the air is so great that this invented story could be literally true. The carbon atom could have become a colorful leaf, a floral perfume, a crab or a fish, a piece of paper in a library archive, or the canvas for a painting.

Finally, imagine that the carbon atom is in a glass of milk (from the air to the grass to the cow and its milk) and you drink it. The carbon atom enters your blood stream and enters a nerve cell that is part of your brain. The carbon atom becomes a part of you who are reading this sentence.

WOW! The atoms from which we are made are recycled throughout all of nature.

Reflection

I knew our bodies were made of carbon, but to think that our brains, the part of us that thinks is made up of the same thing is mind-boggling.

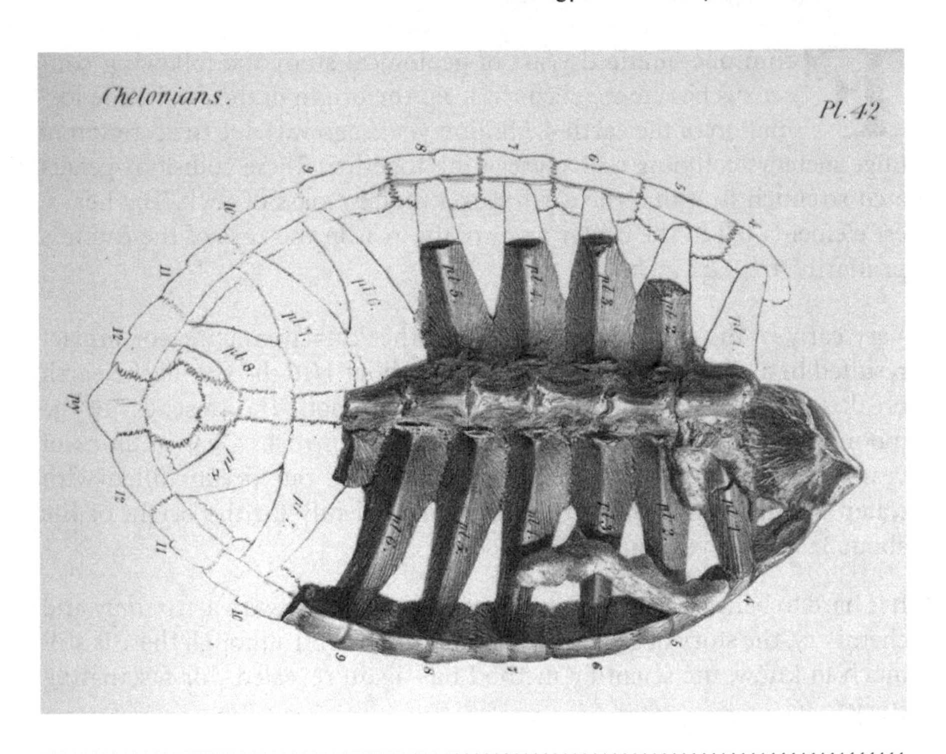

FIGURE 3.1

Above is a fossil of a bone fragment of the underside of a turtle shell. Note how clearly it fits into the structure of the bone as shown by this beautiful illustration.

CHAPTER THREE Geology and Our Dynamic Planet

"Even a medium-sized impact of five kilometers [3.1 miles] across would have turned the oceans to steam, shrouding the planet in enormous, lightning-charged, pitch-black cloud banks that would have lasted thousands of years before they rained out to reform the oceans. Had any life gained a foothold in this hellish setting, it would have been wiped out as the entire planet was sterilized over and over again."[1]

~William Schopf

From one hundred years of geological study the following consensus has emerged concerning the origin of the earth. The formation of the earth 4.5 billion years ago was due to accretion of huge meteors colliding and aggregating together. These collisions generated so much heat that the earth was a molten mass of rock. The heaviest elements fell to the center and are the reason the core of the earth is primarily iron and nickel.

Very early in the origin of our earth another cataclysmic meteor impact resulted in a chunk of the earth's mantle about 1/16 the size of the earth breaking off. This smaller mass became the moon. The closeness of the moon to the earth created huge tides of molten rock. Over millions of years the earth cooled, the moon receded, and our oceans filled with water. Finally, conditions settled down sufficiently for the origin of life about 3.5 billion years ago.

It is hard to believe that we are on the same planet! As with astronomy and chemistry, the story of geology is quite dramatic and although there is still much to know, the scientific method has again revealed this fascinating story to us.

Nils Steensen (1638-1686), also known as Stenno, had some very radical ideas that he published in his *Dissertation on a Solid Body* in 1669. For example, he stated the layers (strata) of the earth constitute a record of earth history over large periods of time and fossils were once living plants and animals. These ideas are true but sad is the fate of many people who have ideas ahead of their time. I imagine the public support for his thinking was nil. After publishing his work, little is known of Stenno except that he became a religious ascetic.

One hundred years later, James Hutton (1726-1797) proposed the idea that Stenno had proposed earlier, namely that natural processes shaped the earth over long periods of time. These long periods were interrupted by sudden catastrophic changes such as earthquakes and volcanic eruptions. Envision a layer of rock spanning hundreds of miles with fossils embedded in it. Suddenly there is a section where the fossils have been uplifted to a height of thousands of feet. Imagine the surprise when a particular fish fossil is discovered atop a mountain. To put it mildly, the earth is quite powerful! Vertebrate fossils were thus discovered and used as signatures for geological strata. The idea of the earth as a dynamic system operating over very long time periods was establishing more credibility.

As a window into the mind of a scientist, consider the following story. It is said that when Hutton found geological evidence that granite is formed from melted matter cooling under high pressure, he experienced such joy and exultation that his guides thought that he must have discovered gold or silver. Instead his elation was because he had observed verification of his idea.

Charles Lyell (1797-1875) published his three volume *Principles of Geology* (1830-1833) in which he wrote the earth was millions of years old. He knew that many people in early nineteenth century England believed the biblical account that the earth was fixed and unchanging, yet he could not escape evidence to the contrary. For example: Mount Vesuvius erupted in southern Italy in 79 CE depositing fifteen to eighteen feet of rock and ash on the city of Pompeii; Helice and Buris (former Greek cities) are now under the sea; the island of Sicily was connected by an isthmus to the rest of Italy; and Pharos, a former island, is now connected to Egypt because of the growth of deltas. There is no doubting the earth to be a dynamic, powerful, and changing planet.

As you think of the geologic past, imagine a different arrangement of both the continents and a different length of day. How do we know? Certain coral fossils provide a window into the prehistoric world. Comparing present day coral to coral fossils we see some interesting differences. In living coral shells, one line is formed every day and each month there is a darker line. The lines in 400 million year old coral fossils are spaced more closely together. This suggests the length of day was only twenty-one hours, implying that the earth was spinning faster. Some think that the earth and the moon were also closer so the level of the tides was much stronger than today (see beginning of chapter). This stronger gravitational influence of the moon on the earth contributed to slowing the rotation of the earth on its axis.

The supercontinent of Pangea separated into the map of continents we know today over the last 100 million years (Figure 3.2). The puzzle of how the continents fit together is solved by similarities of both fossils and geologic strata on the periphery of today's continents. For example, specific species of fern spores, distinctive gray to black shale, and specific magnetic field orientations mark the strata on the East coast of America and on the West coast of North Africa. As the continents migrated apart they developed more unique environments influencing the evolution of plants and animals. When the continent of Antarctica migrated towards the South Pole there was a change in ocean currents in the southern

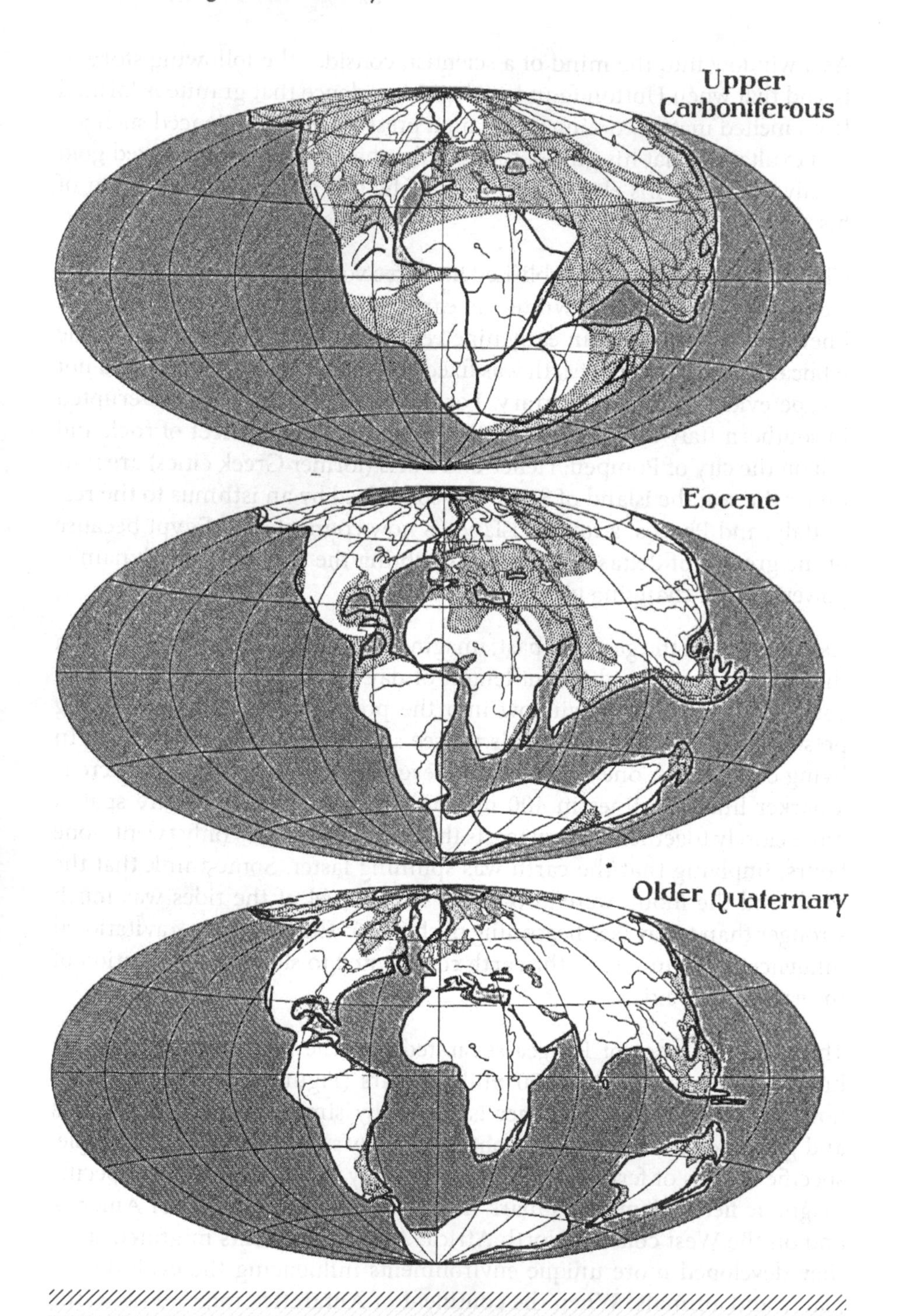

FIGURE 3.2
From the 1915 book by Alfred Wegener diagramming the movement of the continents over the last 100 million years.

hemisphere resulting in the cooling of ocean waters. This environmental change resulted in the evolution of fish uniquely suited for survival in the colder water.

Where does the energy come from that causes the movement of the continents? The answer was uncovered by Harry Hess (1906-1969) when he was serving in the Navy during World War II. He used sonar (equipment that bounced sound waves off objects in the ocean) not only to identify the presence of enemy submarines, but also to map the topography of the ocean floor. He discovered flat-topped mountains (subsequently named the mid-Atlantic ridge) spanning 10,000 miles from north to south under the Atlantic Ocean. Sometimes the mountain ridges reach above sea level, forming islands (for example the Azores). The valley between the ridges results from a slow release of molten rock from the mantle of the earth through a space in the bottom of the Atlantic Ocean. This causes a lateral movement of the ocean floor (like a conveyer belt) spreading both east and west from the ridge. The rate of the spreading is slow, approximately 0.5 to 4 inches per year. The age of the ocean floor at the boundary with the continents is 175 million years indicative of the time it takes for the ocean floor to spread from the mid-Atlantic ridge to meet the continents. The Atlantic grows wider while the Pacific is shrinking. It is the expansion of the ocean floor that drives the movement of the continents. The source of the energy is convection of heat by rising plumes of hot rock deep within the earth. The buildup of pressure at the boundary of the tectonic plates causes earthquakes and tsunamis.

The device that detects earthquakes is called a seismometer and the first seismometer was built in China during the Yan Chia period (132 CE). In addition to its ingenious design, the seismometer provided information about the occurrence of earthquakes far away from their epicenter. Zhang Heng, builder of the first seismometer, achieved fame when he detected an earthquake 1000 km from the capital. Even though this device could not predict earthquakes, the knowledge of the earthquake enabled the emperor to mobilize food and troops to the suffering region as soon as possible.

> *"During the first year of the Yan Chia period (132AD), Zhang Heng constructed the Hou Feng Di Dong Yi. The instrument was cast with bronze. The outer appearance of it was like a jar with a diameter around eight chi (feet). The cover was protruded and it looked like a wine vessel. There were decorations of inscriptions and animals on it. There was a du zhu (a pillar) in the center of the interior and eight*

transmitting rods near the pillar. There were eight dragons attached to the outside of the vessel, facing in the principal directions of the compass. Below each dragon rested a toad with its mouth open toward the dragon. Each dragon's mouth contained a bronze ball. The intricate mechanism used was hidden inside the device. When the ground moved, the ball located favorably to the direction of ground movement would drop out of the dragon's mouth and fall into the mouth of a bronze toad waiting below. The clang would signify that there had been an earthquake. The direction faced by the dragon that had dropped the ball would be the direction from which the shaking came. And each earthquake only made one ball drop. The device worked accurately. The record showed that one time, the dragon spilled a ball but no earthquake was felt. Scholars in the city thought it was odd. Several days later, news came that an earthquake had indeed occurred in the area Longxi. People then realized its ingenuity. From then on, the historian was ordered to record the direction of the quake origins using the device."[2]

~Hong-Sen Yan

For a picture of Zhang Heng's seismoscope go to: inventors.about.com/library/inventors/blseismograph2.htm

How Do We Know?

DATA FOR THE STRUCTURE OF THE EARTH

How did we discover the inner structure of the earth that is about 8000 miles in diameter? Waves triggered by earthquakes have turned out to be tremendously useful. Similar to a CAT scan using X-rays to see inside the body, these seismographic tracings have been used to study the interior composition of the earth. Specifically, the type of seismic wave and its velocity gives us insight into the composition and thickness of the layers of the earth.

Many seismic stations positioned around the globe are equipped to detect seismic waves. Visualize the earth as a sphere. An earthquake is detected at the top of the sphere (Figure 3.3). The shear (S) waves travel on the surface of the earth in both directions and also at all angles through the earth. These waves are detected by seismometers around the globe and the velocities of the seismic waves are determined (Figure 3.4).

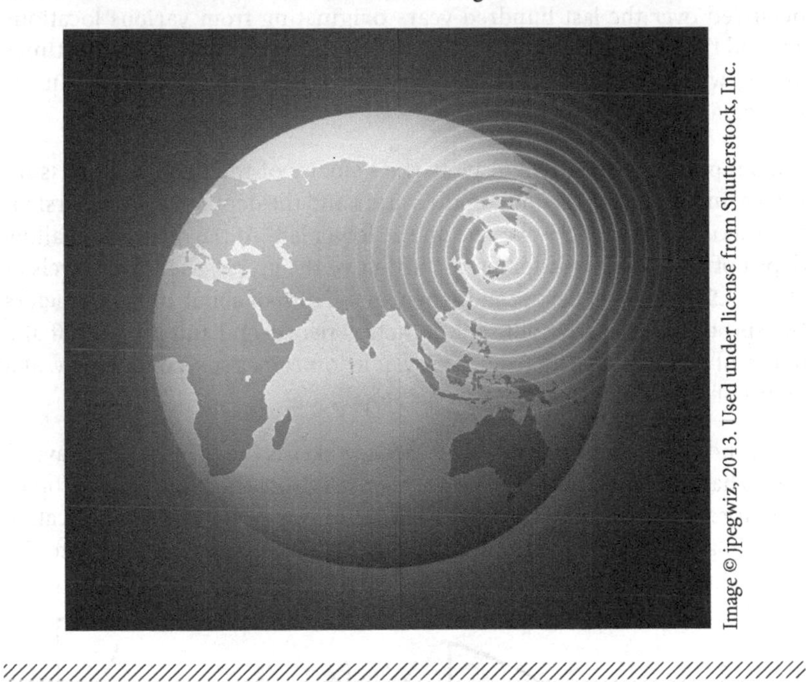

FIGURE 3.3.
Earth with quake waves emanating from Japan.

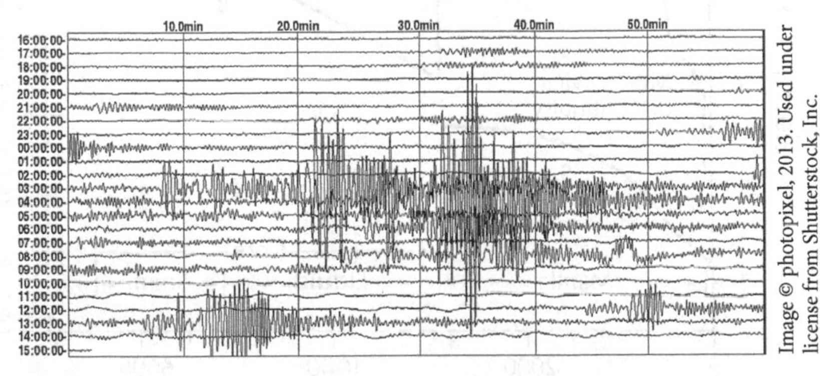

FIGURE 3.4.
Seismic Diagram illustrating seismic waves arriving at different stations.

The S waves are unable to be detected on the other side of the earth (180 degrees) because the center of the earth contains liquid iron and nickel that blocks transmission of the S waves. Thousands of earthquakes have occurred over the last hundred years originating from various locations around the earth so these "experiments" have been conducted many times always with consistent results. (See http://www.iris.edu/hq/ for more information about seismology.)

The graph below (Figure 3.5) plots the velocity of the shear and pressure waves (in kilometers/second) on the y-axis and the depth (in kilometers) on the x-axis. Follow the S wave line on the lower left. At the relatively shallow depth of about twenty kilometers (far left of the graph) the wave travels at about 4.5 kilometers per second. Then, at a depth of about 1000 kilometers the speed increases to about six kilometers per second and it levels off at a depth of 2000 kilometers to about seven kilometers per second. Lastly, at a depth of 3000 kilometers the S waves stop.

A second type of wave triggered by earthquakes is the pressure (P) wave. P waves have different properties from S waves. Note the precipitous drop in velocity at precisely the same spot where the S waves stop. This is indicative of the mantle/core boundary where a granite-like mantle is adjacent to the

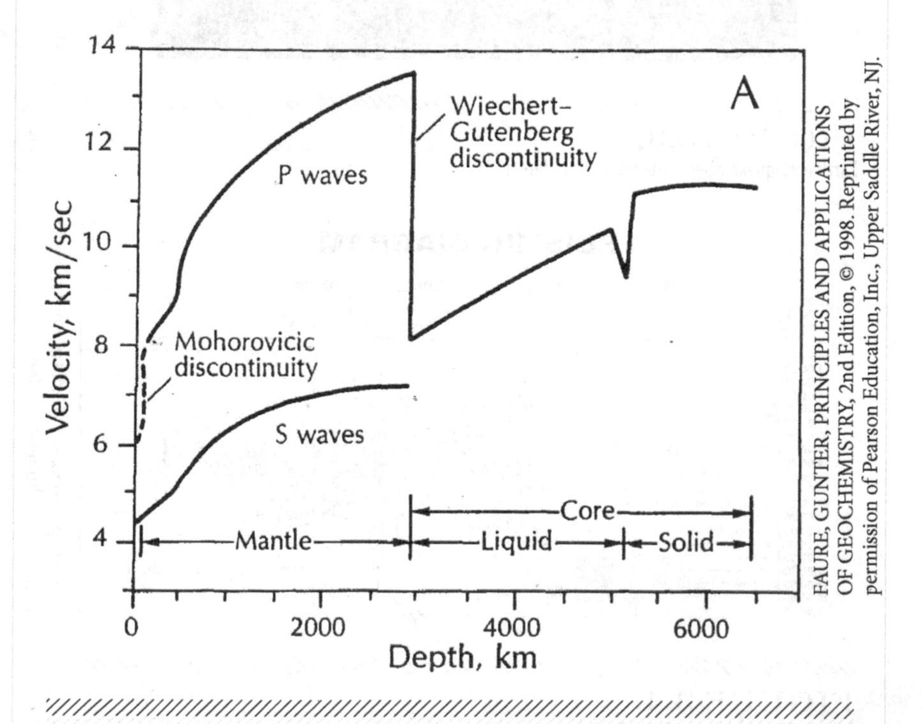

FIGURE 3.5.
Velocity of waves generated by earthquakes vs. depth in the earth.

liquid iron core. The graph also shows a jag in the line at a depth of about 5000 km indicative of the P wave velocity difference between the liquid and solid cores. Changes in the slopes of the lines are named after the scientists who discovered them. Lastly, the velocity of the P wave changes suddenly when it passes from the crust to the mantle. This is called the Mohorovicic discontinuity. This is because the earth's crust has a different composition from the mantle and core. The vertical slope of the line indicates that the crust is only 100 km thick.

When we look at the moon through a telescope we see many craters from meteorites that have impacted its surface. The moon has no atmosphere and no weather, thus the meteors do not burn up. They make long-lived impact craters (Figure 3.6). Luis Alvarez (1911-1988) and others reasoned that the same has also happened on earth although the traces are harder to find. While our atmosphere protects the earth, we are not immune from meteorite impacts. Among many meteorites to impact the earth, we believe one collided with the earth about sixty-five million years ago in the vicinity of the Yucatan Peninsula in Mexico. This meteorite was about six miles across, and when it hurled into the earth it created an explosion equal to millions of nuclear bombs. The result was a cataclysmic environmental change triggered by shock waves and particulate matter thrown into the atmosphere. These dramatic changes in the environment, we believe, drove the dinosaurs into extinction. Small animals with short generation times and large litters (like rodents) are more adaptable to environmental changes and it is from these smaller animals that we evolved. Dinosaurs were large, had long generation times, and laid only one or a few eggs. It is still a mystery that dinosaurs disappeared while other creatures survived that cataclysmic event.

Some meteorites are rich in an element that is rare on earth called iridium. In support of Alvarez's idea, there is a centimeter thick layer of rock containing iridium at many places around the earth in the stratum from sixty-five million years ago. The perimeter of the crater also has diamonds that form only under the very high pressure and temperature characteristic of a meteorite impact. The observations make us confident that a meteor impacted the earth at the time of the extinction of the dinosaurs. About every 100,000 years, the earth is hit by a kilometer-sized meteor triggering nature's equivalent of a nuclear winter. To get a feel for the magnitude of the damage that meteor impacts can cause consider a meteorite seventy yards in diameter that entered the atmosphere in Siberia in 1908.

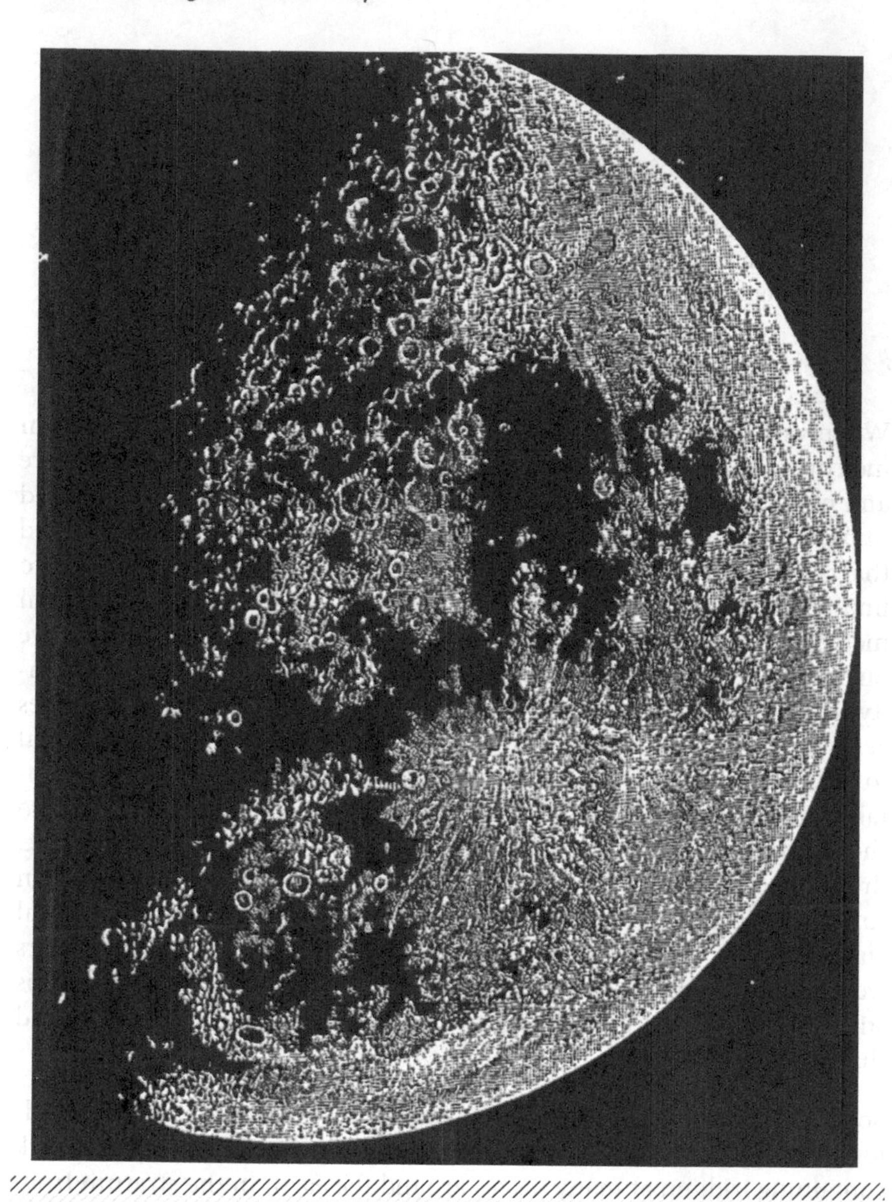

FIGURE 3.6.
Picture of the moon with craters.

It burned up before hitting the ground and the explosion was equivalent to twenty megatons of TNT. Thousands of square miles of Siberian forest were decimated and a man's clothing was ignited sixty-two miles away. This was all from a meteor that was less than 1/16 mile in diameter. Now imagine the effect 100x worse!

The magnetic field around the earth and the ozone layer at the top of earth's atmosphere have made the earth more hospitable for life. The magnetic field originates from the earth spinning on its axis causing the liquid iron to spin in the earth's liquid core. But, the origin of the ozone layer is more complex. Ozone is derived from free oxygen that was formed from life on the earth. Thus, life created our atmosphere. Early life on earth evolved in an atmosphere lacking free oxygen. Only after photo-synthetic bacteria including blue-green bacteria evolved from these more primitive microbes did the atmosphere begin to change. Evidence for the presence of atmospheric oxygen is iron oxide deposited (red rock) in the strata that formed 2.5 billion years ago. Eventually the oxygen level rose to 20%. Finally, the ultraviolet light from the sun caused a chemical reaction where oxygen at the top of the atmosphere was converted into ozone. This acted as a shield blocking out some of the most harmful ultraviolet rays. Thus, the earth became a more hospitable environment for life to evolve and the rates of mutation and evolution slowed.

Reflection

Though life has changed the environment, we have the environment to thank for life.

Lastly, consider insights that bring together the origin of life, evolution of human beings and the study of fossils. Stephen J. Gould gives both a clear and exciting answer to that question. Many paleontologists have noted an explosion of various life forms about 570 million years ago preserved as fossils in a region of Western Canada called the Burgess Shale. Gould's interpretation of the Burgess Shale is that the fossils indicate a large number of dead ends in the course of evolution.

> *"Most significantly, these dead ends suggest that if we 'replay the tape' it would lead evolution down a pathway radically different from the road actually taken. But the consequent differences in outcome do not imply that evolution is senseless, and without meaningful pattern; the divergent route of the replay would be just as interpretable, just as explainable after the fact, as the actual road. But the diversity of possible itineraries does demonstrate that eventual results cannot be*

predicted at the outset. Each step proceeds for cause, but no finale can be specified at the start, and none would ever occur a second time in the same way, because any pathway proceeds through thousands of improbable stages. Alter any early event, ever so slightly and without apparent importance at the time, and evolution cascades into a radically different channel."[3]

If the earth were to evolve life again, then it would almost certainly go down a different path. In other words, there is a significant probability that we would not exist. How does this make you feel? This makes me look at everything in the living world with even more awe and a greater sense of splendor.

Reflection

Overall, I really like the idea that "life does not exist ON earth," but instead they are partners.

FIGURE 4.1

Isidore Geoffrey St Hilaire. Description of new mammals. Paris: Gide, 1841-1861.

CHAPTER FOUR Biology and the Evolution of Humans

"Our breakthrough was the result of 'night science': a stumbling, wandering exploration of the natural world that relies on intuition as much as it does on the cold orderly logic of 'day science.' In today's vastly expanded scientific enterprise, obsessed with impact factors and competition, we will need much more night science to unveil the many mysteries that remain about the workings of organisms." [1]

~François Jacob

François Jacob, Nobel Laureate in Biology and Medicine, made this remark on the process of science in a recent editorial in Science magazine. This is a reminder that determining how various thoughts fit together is often a more random and intuitive process than doing the actual experiments. For you, the process of reading this book may follow this same path. Perhaps you will wake up in the morning with new relationships between ideas in your mind.

The study of biology has certainly come a long way from the time of Hippocrates (460 BCE - 370 BCE) and Pliny the Elder (23 CE-79 CE). For example, Hippocrates believed that different temperatures between the seasons in Europe caused differences in the quality of semen. He thought the semen of the cold months was different from the semen of the warm months so the offspring looked different from each other depending on the season in which they were born. In contrast, he concluded that humans from Egypt and Libya looked more similar to each other because of the less varied warm climate. He also attributed the calm nature of the Egyptians and Libyans to the warm climate. [2] Pliny the Elder made no distinction between fantasy animals, like the unicorn, and real animals. How could this be? There is a story that Marco Polo, in the thirteenth century, saw animals with a single horn on their muzzle on the island of Java. The animals were black with spiked tongues and heads like a wild boar. The "unicorns" were in fact rhinoceros but Polo interpreted what he saw in the context of what he thought.[3] Over the last 1000 years the details of biology have certainly changed, but even today what we see is strongly colored by what we expect to see.

As with the other sciences, in the seventeenth century, activity began to pick up in the study of biology. Anton von Leuwenhoek (1632-1723), inventor of the microscope, extended his senses to the cellular level and discovered spermatozoa in human and animal semen. At about the same time Robert Hooke (1635-1703) described cork under the microscope as containing "cells." One would think that these independent observations would help to spread the information about cells, but, again, it took more than 150 years until Matthias Schleiden (1804-1881) and Theodore Schwann (1810-1882) proposed cell theory; all cells come from previously existing cells. It was almost immediately and universally accepted.

In the 1600s several other insights into biology were made hundreds of years before acceptance by the scientific community. One concerned the idea of spontaneous generation and the other concerned fossils and the origin of species.

People observed that mice emerged from rotting hay and maggots emerged from rotting meat. Why? Perhaps these life forms generated spontaneously from the hay or the meat? Experiments by Francesco Redi (1627-1697) tested the idea of spontaneous generation by comparing the meat in open flasks to meat in covered flasks. He saw wormy meat only in the open flask and also noted that when the flask was covered with gauze there were never any worms on the meat. This was because the flies couldn't lay their eggs on the meat although they were all around the frame. Despite sharing his evidence that maggots came from fly eggs, thus disproving spontaneous generation, the idea persisted for another 200 years until Louis Pasteur (1822-1895) performed his famous swan-necked flask experiment. The bacteria were prevented from entering a sterile flask with culture liquid by an S-shaped tube on top of the flask. Bacteria from the air were trapped in the curved part of the glassware and the culture liquid remained clear indicating no bacterial growth. Only after the glass swan-neck was broken allowing bacteria from the air to enter the flask did bacteria grow. Thus, spontaneous generation was finally disproven.

John Ray (1627-1705) published in 1693 that the stability of species was not absolute and that fossils were petrified remains of extinct creatures. This was 170 years before Darwin's origin of species and 100 years before this idea of fossils was accepted. Ray did not suggest a mechanism for the origin of species but he definitely set the stage for the concept of evolution.

In the world of biology, Darwin, Wallace, Schleiden, Schwann, Lyell and others all had the good fortune to work in the nineteenth century. This was a time when science, particularly outside of the places controlled by the Church of Rome, was much more respected and new ideas were more easily accepted by society.

Charles Darwin had the first volume of Lyell's newly published *Principles of Geology* with him on his five-year journey on the HMS Beagle (1831-1836). Thus he most likely believed the earth to be millions of years old (time enough for biological evolution). It was during this trip that Darwin visited the Galapagos Islands and among his observations were fourteen different species of finches. They were so tame that to catch them he simply picked them up. They had no reason to evolve a fear of predators and fly away. He noted that the beaks of the finches were perfectly suited to the different environmental conditions on each island. Darwin knew of Lamarck's thinking that species could change over time, although he disagreed with Lamarck's ideas (later discredited) of the inheritance of acquired characteristics. He also read Malthus' essay on *The Principle of Populations* (1798) that stated that populations increase exponentially (2,4,8,16,32,64,etc) whereas food supplies can only increase arithmetically (1,2,3,4,5, etc). Collectively, the work of Lyell, Lamarck, and Malthus coupled with his own observations triggered the thoughts for Darwin that more organisms are born than could ever populate the earth. He reasoned that there must be some form of natural selection and survival of the fittest. I suspect he was practicing "...a stumbling, wandering exploration of the natural world that relies on intuition as much as it does on the cold orderly logic" (from Jacob quote at the beginning of this chapter).

Darwin published *Journal of the Voyage of the Beagle* in 1839, describing his observations, but did not provide a general theory to explain them. During the next twenty years he came up with the theory of the origin of species by natural selection. Then Alfred Wallace contacted Darwin and said that he had come to the same conclusion. Wallace and Darwin co-authored a paper in 1858 titled *"On the Tendency of Species to Form Varieties; and on the perpetuation of varieties and species by Natural Selection"*. And in 1859 Darwin published *On the Origin of Species by Means of Natural Selection*. Although his book created much controversy, it was an immediate success. Later, in 1871, Darwin published *The Descent of Man, and Selection in Relation to Sex* in which he stated:

> *"As many more of each species are born than can possibly survive, and as, consequently, there is a frequently recurring struggle for existence, it follows that any being, if it vary however slightly in a manner profitable to itself under the complex and sometimes varying conditions*

of life, will have a better chance of surviving, and thus be naturally selected. From the strong principle of inheritance, any selected variety will tend to propagate its new and modified form."[4]

Much is known about evolution but many questions remain about the origin of life. When contemporary biologists think about the origin of life they focus on the overlap between biology and chemistry. There is a tremendous amount of chemical information known detailing how cells work. This remarkable complexity is present in every cell on earth, whether it is a bacterial cell, plant cell, or animal cell.

Reflection

It's amazing to think of the vastness of the universe, both in space and time, and of how many billions of things had to go exactly right for us to exist. So many millions of years for one cell to survive – yet it led to all the life around us.

How Do We Know?

FROM CHEMISTRY TO BIOLOGY

Chemistry inside cells causes them to live because chemical reactions enable metabolism to occur. Metabolism provides all of the chemical energy necessary for life and is formed inside cells. The ability to move our muscles, breathe air, see, hear, smell, and even think are all dependent upon chemical energy. Cells comprise every part of our bodies, and food provides the raw materials that ultimately fuels metabolism.

In order for metabolism to occur, two components are needed: substrates and enzymes. The substrates are chemicals like sugar, phosphate, amino acids, fats, and water and these components come from our diet. The enzymes enable conversion of the substrates into products that ultimately generate energy for cells. Without enzymes the chemical reactions necessary for life will not happen fast enough for life to occur. Most enzymes are proteins that are composed of sequences of amino acids.

There are twenty different amino acids and the lengths of proteins range from about twenty-five to 1000 amino acids. The function of an enzyme is determined by its amino acid sequence in the same way that the meaning of a word is determined by its sequence of letters. But proteins are more than a linear sequence of amino acids. Inside cells they fold into 3-dimensional structures (think of paper folding into an origami bird). Each protein structure enables a specific function to be performed. It takes about twenty steps in order to harvest the energy from sugar, and each step requires an enzyme.

In every cell studied, the chemistry of converting sugar into energy is the same. The enzymes in every mammal are similar to enzymes in amphibians, birds, plants, fungi, and even bacteria. As diverse as the biological world looks to our eyes, there is an underlying unity at the chemical level. This underlying sameness is the reason why cells are alive in the first place.

Whenever one cell divides into two it must pass on the ability of each daughter cell to perform its own metabolism. Without it the cell will die. There is a set of directions that specifies how each protein is made, which is coded in the DNA. Each cell is capable of reading the DNA code and translating the code into specific proteins. The proteins then function to create the metabolism necessary for life. As a cell divides it is critical for the DNA to be transferred to each daughter cell, because then each cell can create its own metabolism.

When cells divide they must make copies of their DNA using a process called replication. Cells replicate their DNA in a process that is better than 99.9% correct but not 100% correct. Errors incorporated into the DNA every time a cell divides are called mutations. Evolution is the process of mutation followed by variation and selection. One consequence of mutations may be that nothing changes in the proteins of daughter cells. A second possibility is that the daughter cell dies because the gene no longer codes for a functioning protein. The third possibility is that the daughter cell lives more efficiently than the "mother" cell and a new variety is produced. This cell will better compete in the environment and become the most prevalent cell over time. Mutations in the DNA enable heritable changes to occur which are stored and saved in future generations of cells. These incremental changes in genes are accumulated and unique groups of genes are selected in the environment. This is what drives evolution.

The experiments of Miller and Urey in the early 1950s demonstrated that amino acids and other biological molecules (like DNA bases and sugars) could be made in conditions similar to what existed in the very early earth. They heated and sparked a reaction composed of gases that we think were present in the primitive atmosphere (carbon dioxide, ammonia, methane, and water) and after a week amino acids formed in their flasks. Based on this result, environmental chemistry could have created the chemicals necessary for life before life actually began. DNA bases, glucose, and lipids have also been formed from experiments simulating the early earth. These chemicals may have originated in a shallow pond and as the pond evaporated the chemicals became more concentrated. One idea is that after a rain the lipids spontaneously formed spheres in which the other necessary bio-chemicals (sugar, DNA bases, and amino acids) may have been trapped. These chemicals ultimately formed DNA and proteins necessary for life.

Among the unanswered questions is how the first sequence of DNA came into existence. A long sequence of specific DNA bases had to be assembled in order to provide a genetic code for making proteins. This was a critical early step in the origin of life and gradual modifications in the gene (mutations) were inherited by daughter cells. RNA (a molecule similar to DNA) may have preceded DNA as the genetic material in early forms of life as it may have provided both functions of replication and metabolism.

We don't know what happened at the origin of life but there may have been several steps .[5] Step one may have been a cell that was capable of performing metabolism so the cell could make necessary nutrients and energy to live. Step two may have been a cell capable of replicating long stretches of DNA (or RNA). The final step was the coming together of metabolism and replication creating a cell capable of performing both. A parasite, for example, the cell capable of replication, may have infected another cell, for example, the metabolism cell, and they became symbionts --cells living together for the mutual benefit of both.

Another theory is that life originated in deep oceanic hydrothermal vents. Iron and sulfur could have functioned as catalysts increasing the rates of critical chemical reactions in the absence of proteins. Support for this idea is that bacteria have recently been discovered which thrive in the conditions of these very hot thermal vents and also contain DNA and proteins.

Lastly, life may have come to earth from another planet. Cellular-like structures are seen in the meteorites and these rocks contain amino acids and other biological molecules. Even if this is true, it doesn't solve the problem of the origin of life; it just puts it off to another planet.

One thing we know for sure is that the origin of life happened, because we are here. It is possible that the origin of life happened many times, but under the harsh conditions of the early earth only one cell survived from which we are all descended. Again, we think this because the chemical processes among all cells on earth are overwhelmingly similar.

After the process of photosynthesis evolved, the atmosphere was transformed into one containing oxygen. In the presence of oxygen many early cells died, because of oxygen toxicity, but those that survived were able to use the oxygen effectively, creating more energetic forms of life. The evolution of sexual reproduction occurred after cells adapted to oxygen. This provided a mechanism for shuffling genes for the next generation so individual adaptations could be mixed together into new life forms. The most successful unique combinations of genes developed into organisms that were selected for by the environment. This resulted in plants and animals that could respond to challenges in their environment most effectively and that could reproduce. This was evolution in action. The bounty of new life formed about 500 million years ago is referred to as the Cambrian explosion. During a time as short as ten million years (from a geological perspective) many new classes of organisms (for example; fish, plants, shellfish, and worms) evolved.

You will find the metaphor of the hairpin useful when thinking about evolution. Envision two closely related species each represented by the two ends of the hairpin. The evolutionary relationship between those animals is not that one animal evolved into the other – rather each animal evolved from a common ancestor. The most recent common ancestor is represented by the curved end of the hairpin. Even though the animals look similar, they evolved from a common ancestor many years earlier – represented by the length of the hairpin.[6]

After the evolution of plants and animals in the sea some amphibians migrated onto the land. It was from a common lizard-like ancestor that the dinosaurs evolved about 250 million years ago. When these animals first came onto land, their legs came out of their pelvis to the sides requiring them to sway from side to side when they walked. This swaying motion compressed one lung at a time. For this reason, only one lung at a time functioned when the lizard moved. Consider a mutation (a change in the DNA sequence) resulting in a slightly different structure for the pelvic bone where the legs enter the pelvis from below rather than from the side. The dinosaur now walks upright and both lungs work simultaneously. This is a clear example of a mutation that would be selected for and make a positive influence on the life of the organism through evolution. The age of dinosaurs spanned from 250 to 65 million years ago.

Small animals eat less and reproduce more quickly than larger animals, thus they can survive environmental disturbances more readily than larger animals. As a result of the extinction of the dinosaurs, ecological niches opened up for new species to inhabit. As the dinosaurs decreased in numbers, small mammals flourished; it was from these mammals that humans evolved.

Several methods are used to determine evolutionary relationships between organisms. They are the DNA sequences of various organisms; similarities of bone structures and physiology; biochemical and genetic similarities. All of these data point to the same evolutionary tree. We evolved from common ancestors of other mammals over the last eighty million years.

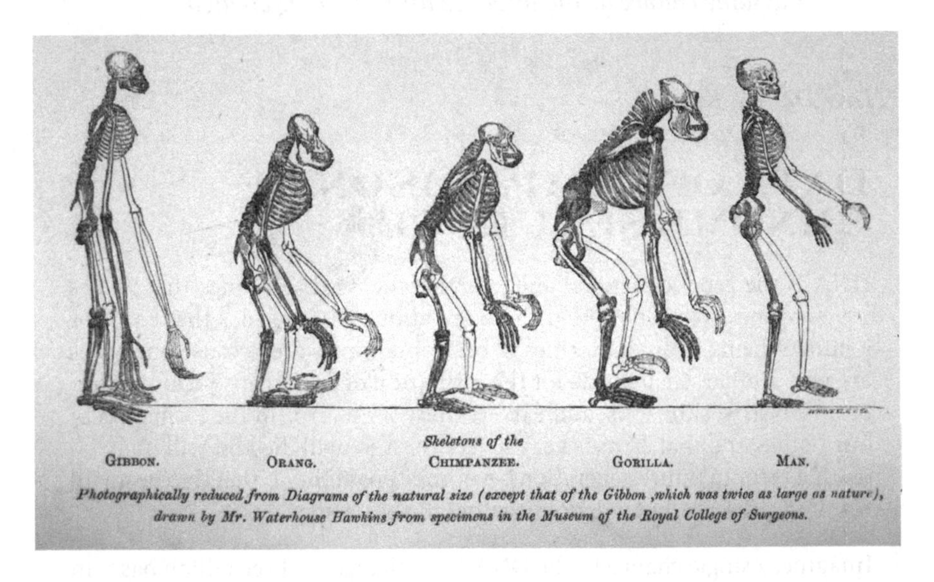
Skeletons of the
Gibbon.
Orang.
Chimpanzee.
Gorilla.
Man.
Photographically reduced from Diagrams of the natural size (except that of the Gibbon, which was twice as large as nature), drawn by Mr. Waterhouse Hawkins from specimens in the Museum of the Royal College of Surgeons.

FIGURE 4.2

"Photographically reduced from Diagrams of the natural size (except that of the Gibbon, which was twice as large as nature), drawn by Mr. Waterhouse Hawkins from specimens in the Museum of the Royal College of Surgeons."

> "Nothing in Biology makes sense except in the light of evolution."[7]
> ~Theodosius Dobzhansky

Reflections

It still blows my mind that we as humans are all related back to animals. To us, one week seems to stretch on forever, but thinking about how long it took the earth and other planets and sun to form after the explosion of the first supernova, or how long it took for humans to evolve from the first cells, a week seems comparable to the blink of an eye. Knowing what I know now, I will never look at the stars the same way again.

Even though it took nine months to create you, it took millions and millions of years of mutations and evolution to create humans.

It's crazy how something as little as winning the lottery has the same chance as the universe and life being created.

How Do We Know?

DATA OF COMPARISON OF GENOME SEQUENCES

DNA is the genetic material and it is the chemical substance that makes up our genes. We believe that there are about 22,000 genes that code for a human being. Our similarities to other mammals are because our genes are very similar. Genes code for the formation of our organs including our brains, hearts, skin, eyes, and ears. If there is an error in the formation of our bodies and that error is very severe then we will die and will not pass on that error to future generations. Another possibility is that the error will result in a disease, but will not be lethal to the person.

Imagine, a single change in the DNA (one change in three billion bases in the DNA) can result in a male appearing more like a female! This is because the functioning of the male sex hormone, testosterone, is blocked. I particularly like this story because it was brought to me by a student and she was empowering herself to learn. She didn't know where the information would lead her and, in a sense, she went on a journey. It was a journey of thought, research, and learning.

Below is the data showing that the gene that codes for the testosterone receptor can account for this problem (Figure 4.3). About twenty-five years ago, getting the DNA sequence data represented by these different peaks may have taken months of hard work. The result would be a cause for celebration for the researcher as it represented the fruit of a lot of hard work and dedication. Such great progress has occurred in the last twenty-five years that rather than months of work this result may have been obtained in as little as one day! Therefore, now a different set of questions can be asked since getting the gene sequence is so easy and routine. For example, how evolutionarily conserved is this gene? Our reason for asking this question is that we have an assumption that if the gene is very important to the survival of the animal then it will be conserved during evolution. The methods for analyzing genes from various organisms enabled the following figure to be assembled. It shows the incredible similarity of this gene among all mammals from humans down to rats! You may not find that a pleasant thought but it is true.

The vertical line at position L830 is always the letter L (which is the symbol for a particular amino acid named leucine). This particular position is always an L when we compare the amino acid sequences of this gene from humans to mice. This underscores the importance of that particular amino acid in the position 830 of the protein. Changes in that spot resulted in an animal that was unable to reproduce and so that change was not passed on. Over the last one hundred million years there have been numerous mutations in many different genes. These changes may have resulted in the death of the animal but on occasion they resulted in small changes that led to the development of new variants of the organisms and ultimately to the evolution of new species.

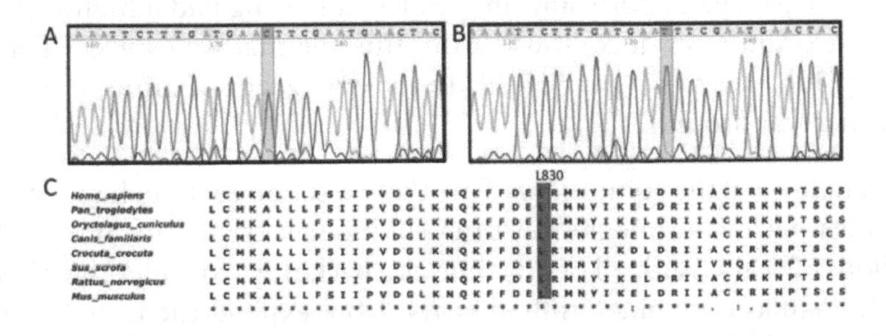

FIGURE 4.3

The single change in a base of DNA results in a change from one amino acid to another in a particular protein. Specifically, the amino acid leucine (L) changed to the amino acid phenylalanine (F) and that accounted for the disease -- androgen insensitivity syndrome. This resulted in failure of normal male external genitalia even though the individual is genetically male (XY). Panel C shows species that have very similar amino acid sequences (in order): human, chimpanzee, rabbit, dog, spotted hyena, wild boar, brown rat, and house mouse.

The single letters in panel C represent the sequence of amino acids in the particular protein. The sequence of amino acids determines the meaning of the protein and if the protein does not have the correct sequence of amino acids then it will not function properly. For this specific gene, a single protein is made that does not function properly resulting in feminization in a male. A single chemical change in the DNA from C (top left) to T (top right) accounts for the disease (see shaded region). If you are a healthy person, be thankful that all of your genes are working properly!

There are certainly similarities between humans and other animals but what are equally – if not more – fascinating are our differences. Biologists have learned that there are genes associated with development of the nervous system and brain that are different between humans and other primates. How do these genes result in human consciousness? As much as we know there is much more yet to know.

Here is an interesting story speculating about the importance of song in the development of consciousness by Daniel Hillis. About two million years ago there was a group of apes able to mimic each other's sounds. Let's call these sounds songs. These songs were "passed on" from mother to child and certain songs were more common than others among members of the group of apes. In a sense, one song survived better than other songs. As the apes survived they reproduced and expanded the songs as part of their lives. Over time, certain of these songs became specialized to critical events in the life of the ape like food, mating, or danger. The ape that was best able to both sing and understand song had a higher chance to survive and reproduce. It was from this ape that we evolved. For this reason Hillis has said, "music leads to the emergence of mind rather than the other way around." [8]

How could all of the complexity of life that we see on earth to have evolved? A critical factor is time and fortunately life has existed for hundreds of millions of years. With all of this time as a backdrop, consider the life span of a human as only one hundred years. Now, expand the time to 10,000 lifetimes and we are to one million years. If the probability that something happens in your lifetime is one in a million then it will probably not happen, but in one million years that event happens with odds of about three to two in favor. Each of these one-in-a-million events becomes cumulative because of the mechanism of biological evolution. This is why it is critical that life has existed for millions of years because it has taken that long for life to evolve. [9] Quite remarkable!

As you have read from astronomy, chemistry, and geology to biology it seems that humans have been reduced to a pointless accident in an indifferent universe. But Marcelo Gleiser, professor of philosophy, physics and astronomy at Dartmouth College, has argued "we are a rare accident and thus not pointless." He reminds us of the difference between life and intelligent life. The appearance of bacteria and algae filling every niche is probably not unique to earth. But single celled life is not multicellular and intelligent life. This is much more rare. Recall that earth has a long-lived sun and contains a protective oxygen-rich atmosphere. Our ozone layer and magnetic field protect plants and animals on the surface from cosmic and ultraviolet radiation. Our sun steadily releases just the right amount of energy to sustain us. For these reasons, the complexity of life on earth and our own existence are most likely exceedingly rare in the universe. Even if the Search for Extraterrestrial Intelligence (SETI) does find intelligent life elsewhere in the universe it will be immensely far away. Gleiser ends by saying:

> *"...And if we are alone, and alone are aware of what it means to be alive and of the importance of remaining alive, we gain a new kind of cosmic centrality, very different and much more meaningful than the religion-inspired one of pre-Copernican days, when Earth was the center of Creation. We matter because we are rare and we know it.*
>
> *The joint realization that we live in a remarkable cosmic cocoon and can create languages and rocket ships in an otherwise apparently dumb universe ought to be transformative. Until we find other self-aware intelligences, we are how the universe thinks. We might as well start enjoying one another's company."*[10]

Reflection

Science encompasses numerous studies, from astronomy, the study of the universe beyond earth, to chemistry, the study of elementary forms of matter and everything in between. Each of these studies attempts to answer questions regarding why things happen the way that they do, what causes various reactions, and even, in their own ways, to answer how our universe began and how life originated. What have we found with this expansion in our ability as scientists to uncover truths about our universe? We do not know much at all. With each theory that we have proved or disproved we have found more questions to ask. Also, as technology has grown our ability to test theories has increased so that we are able to think more in depth in all areas of science. So yes, we are asking questions and finding answers, however, when we do so we also realize how expansive our universe is, and as we have recently learned it is constantly expanding at an accelerating rate. What we really know is that there is so much more to know.

Image © Eugenio Marongiu, 2013. Used under license from Shutterstock, Inc.

FIGURE 5.1

The devastation in this picture elicits the feeling of solastalgia (a word coined in 2003 by Glenn Albrecht, an Australian environmental scientist and philosopher). Solastalgia is a combination of solacium (Latin for comfort) and the Greek root –algia (pain) and means the distress of environmental change on people in their home environment.

CHAPTER FIVE Reflections on Science Literacy and Sustainability

"One thing I have learned in a long life: that all our science, measured against reality, is primitive and childlike - and yet it is the most precious thing we have."[1]

~Albert Einstein

I'd like to begin this chapter on a personal note. I recently had to put my dog to sleep after he'd been with my family for eleven years. At one moment he was there and then -- he was gone. His body was there but he was gone. His body quickly grew very cold. The same is true for people. At one moment we are alive and then we are gone. At that moment, the essence of each one of us leaves. What is that essence within each of us? As much as we have learned from science, there is certainly much mystery that remains to be learned and understood. These questions may be beyond science. But, there is much that we do know that needs to be celebrated as a significant accomplishment of human minds.

You have made it through the scientific story of the universe and the origin of life spanning billions of years. Along the way you have been through the incredible vastness and coldness of our incessantly expanding universe. You have learned about our chemistry and the similarity between atoms on earth and in the most distant galaxies. We believe the chemistry of the universe is the same everywhere and it is chemistry from which all life is made. The earth's incredible power is difficult to see (unless we are experiencing an earthquake or volcanic explosion) given the slow pace and immensity of geological time. The one-in-a-million events that very rarely occur in a lifetime become almost certainties in a million years and that is the time scale for biological evolution. Science, through its method of observation, hypothesis, experiment, and conclusion, carried out over a period of 400 or more years has resulted in this story of the universe and our origins.

We are on a tiny planet (earth) racing and spinning through space in a huge universe that is expanding at the rate of thousands of miles per second. How does that make you feel? Insignificant? Meaningless? Tiny? So shall we reject the scientific facts about our world because they make us uncomfortable? Or perhaps we can think about the awesomeness of the universe and even though there are many unanswered questions, we can appreciate our ability to understand it at all. Recall, our own rarity and uniqueness (to the best of our knowledge) is how the universe thinks! The quotes from astronauts returning from space say that our earth is like the most precious pearl in the blackness and void of space.

> "*...The earth was small, light blue, and so touchingly alone, our home that must be defended like a holy relic...*"[2]
>
> (Aleksei Leonov, cosmonaut)

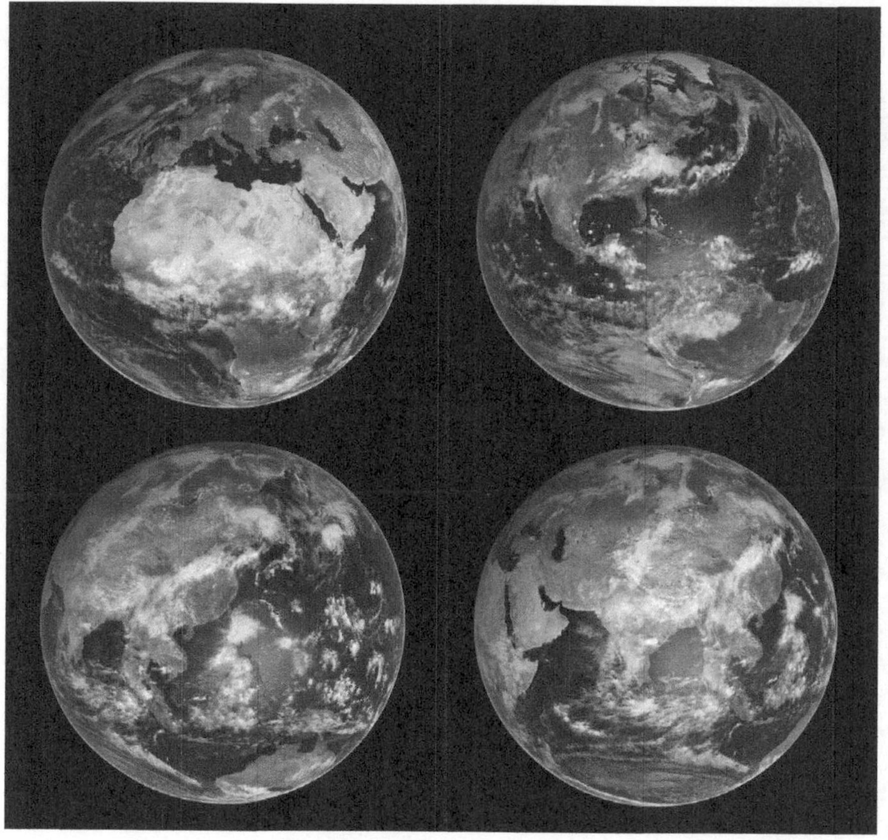

FIGURE 5.2
Pictures of Earth from Space.

Consider the interdependence of all life on each other and the environment. Let's zoom in on one acre of land and, what do we see? There is soil as a layer on which plants are growing. Then, there is a layer of insects followed by a layer of birds and rodents. The layers culminate with the large carnivores at the top. Visualize this as a pyramid. Energy from the sun flows through this pyramid from the bottom up because it is the plants that absorb the sunlight as they take in carbon dioxide from the air and release oxygen. Each layer can also be thought of as a link in a food chain. For example, the soil supports an oak tree, a leaf of which is eaten by a deer that is killed and eaten by a hunter. But this is just one example from many; there are also insects and birds that are dependent upon that oak tree. So, there is a tangle of food chains in our pyramid: one food chain is soil-oak tree-deer-human and another is soil-oak tree-insect-bird. Envision that there are thousands of overlapping food chains in every environment. The myriad of interacting species of plants, insects, and animals (as well as bacteria although we cannot see them) leads to both ecological complexity and ecological stability.

So, what is sustainability and why should we behave sustainably? It is more than recycling and planting trees. It is a mindset linking our understanding of the global ecology of the natural world with good planning for the future. The science-based story of the universe and our origins leaves no doubt that we are intimately part of nature and the environment is part of us. Our burning of fossil fuels (coal, gas, and oil) sustains our energy-rich life styles and increases the carbon dioxide in the atmosphere. Even though this is hard to believe -- we are affecting the global climate. This is the consensus of greater than 90% of the scientists. Chemicals in the air, water, and food that we eat also affect human health, underscoring the relationship between medicine and the environment. This also underscores the relationship between environmental health and human health.

During the early 1970s there were major efforts by the United Nations to reduce population growth, poverty, and degradation of the environment. The momentum generated from this effort gave rise to many initiatives, among them the Environmental Protection Agency, and The Clean Air and Clean Water Acts. In some cases, water quality has improved as in the success story of the Hudson River that has become restored over the last twenty-five years. But many challenges remain. Lester Brown, founder of the Earth Policy Institute, has recently written *Full Planet, Empty Plates*, focusing on our problems of overpopulation and food scarcity.

The pace of change is so rapid that we must find ways to implement good ideas more quickly now than in the past. The years that often pass between a first discovery and its acceptance by society are too long. How can we increase the pace of assimilation of ideas into culture? In this era of the Internet we can share ideas among millions of people. Let's also share the data with clear descriptions of specific experiments and educate a science literate population. Then, people can decide for themselves what to believe. In *The Sun, The Genome and The Internet*, Freeman Dyson discusses why the work of Alfred Wegener was ignored for so long. Wegener presented evidence in 1915 about continental drift, but his evidence contradicted prevailing dogma. There was no conceptual framework for understanding how continents could move. This question was answered about thirty years later (see p. 41) and the concept of continental drift became accepted very quickly thereafter. As an increasing percentage of society becomes science literate, I predict ideas supported with data will quickly gain societal acceptance and traction. This is particularly critical on the issues of the environment and sustainability.

In order to answer questions for ourselves we need to remember the difference between observation and inference. Observation is what we see. Inference is what we believe the observations mean. As people become more science literate, we will improve our own observations and our own inferences. We will become more confident in asking, "How do you know ...?" This mindset will increase the number of people making decisions for themselves based on evidence.

Approaching questions scientifically presents an opportunity to put our own prejudices and biases to the side and begin with careful observation (the first step in the scientific method). In this sense, science represents a method for bringing people of disparate points of view together. If there is a difference over what people observe, then let it be openly discussed. Similarly, if there is a difference over what question(s) to ask, then let them be debated. By doing so, we will build more productive, sensitive and responsible relationships with each other. From these relationships come more sustainable, aware, and compassionate communities.

Reflection

Does it matter that we know the Story of the Universe? The story allows us to appreciate the uniqueness of the earth. This appreciation leads to study and research. Research leads to data and understanding. Understanding leads to predictions. Predictions can lead to present-day action.

Since 1958, the level of atmospheric carbon dioxide has been measured from the top of the Mauna Loa dormant volcano on the big island of Hawaii (Figure 5.3). There has been a steady increase from 315 ppm (parts per million) in 1958 to 385 ppm in 2008. This is primarily due to the burning of fossil fuels and is contributing to global climate change. In addition to the steady increase, the line has regular wiggles due to the seasonal growth and decay of vegetation.

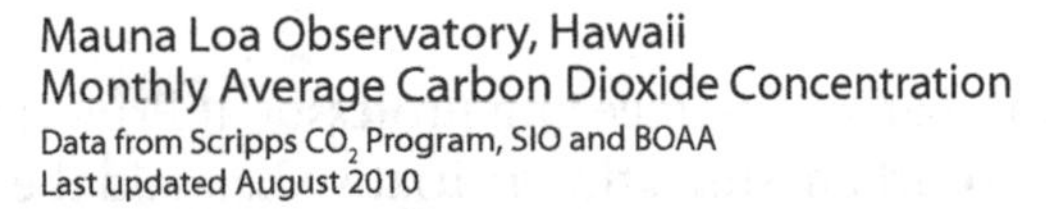

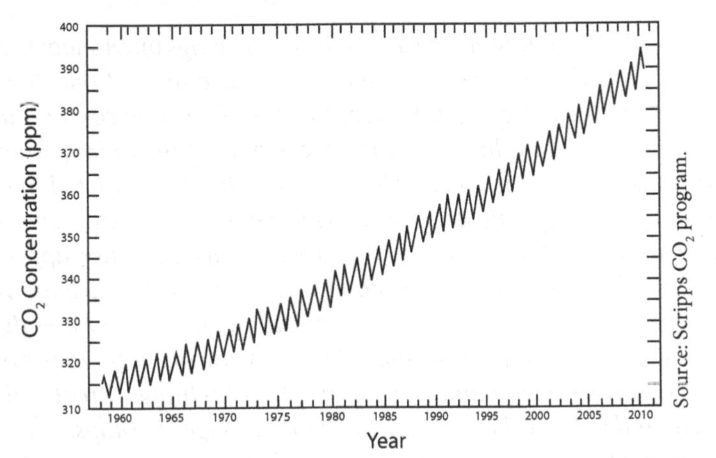

FIGURE 5.3

Increase in atmospheric carbon dioxide between 1958 and 2008.

I want you now to think about the future. Think about the state of the world with its changing ecology, wars, poverty, and hunger and also with pockets of beauty, love, and harmony. In what direction do you think the world is heading? Now consider the metaphor of a train that is carrying all of us forward in time.

Who is driving?

How fast is it going?

Where are we going?

What will we do when we get there?

Are we there yet?

We are too busy to concern ourselves with these questions in our daily lives but they are certainly critical. The Universe Story (from astronomy through chemistry, geology, and biology) makes very clear that all things in our world are interconnected and we all come out of the same pot. Here I use the word "pot" to refer to all elements of the environment on which we depend -- the sun, the water, the tides, the weather, the earth, the air, the food, all of the plants and animals, all of the bacteria, and many other things. That pot better be clean and beautiful for the benefit of us and everything else!

Rob Socolow, a mechanical engineering professor at Princeton University, who also leads the Carbon Mitigation Initiative has said the following:

> *"Confronted with unwelcome news, human beings often shoot the messenger. Consider two earlier occasions. Galileo argued that the earth wasn't at the center of the universe. For this, he was excommunicated. Darwin argued that human beings were part of the animal kingdom, and he was cruelly mocked. The idea that humans can't change our planet is as out-of-date and wrong as the earth-centered universe and the separate creation of man, but all three ideas have such appeal that they will fade away only very slowly ... we should anticipate robust resistance to the message that we are fouling our own nest with fossil fuel emissions and deforestation. Armed with insights from psychology and history, communicators of the climate change threat will more deeply understand the hostility to their message. Perhaps, communication will be more effective when shared concerns are acknowledged.*[3]

Socolow has developed the concept of the stabilization triangle. Visualize two lines comparing carbon emissions: one line represents the world behaving with no change from today for the next fifty years and a second line represents a new strategy implementing new climate friendly technologies that emit less greenhouse gasses. (The graph would be similar to Figure 5.3, above, but two lines would have different slopes.) Strategies that reduce the slope of the lines represent measureable efforts to mitigate climate change. Several fifty-year strategies include electricity end-use efficiency, passenger vehicle efficiency, wind power, solar power, population controls, etc. We are beyond the carrying capacity of the earth. This means that both our lifestyles and population size are not sustainable with the resources of the earth. I believe this information is being ignored in part because of the lack of science literacy among our population.

The climate change effort must be balanced with efforts to protect water resources and biodiversity as well as prevent nuclear war and decrease global poverty. Fortunately, as we reduce global poverty we have experienced reductions in the population growth rate.

The critical challenge now is how to translate knowledge into action. Perhaps the following story from neuroscience will help to point the way. Consider your brain and the images that flood into it every instant of your lives. Images flow into your brain from your eyes as well as inputs from all of your other senses (touch, smell, taste, sound) and send a huge amount of information into your brain. For myself, this morning I got up, shaved and showered, got dressed, walked the dog, cooked and ate breakfast, drove to work, etc. Fortunately, I forget most of the information that I receive, or I would still be thinking of the traffic light from 9:00 this morning! What about you?

With that as background -- what is it that enables you to remember certain events in your lives? Where were you on 9/11? Where were you when you first fell in love? These events are emotional and research shows that we tend to remember events better when they are coupled to our emotions. Our memory is tied to our emotions. We remember certain events most vividly and these add meaning to our lives.

Remember solastalgia? (Figure 5.1) It means the distress individuals experience when their home environment is being degraded. No matter what we do, the earth will continue to evolve in her own way. My worry is that human time scales are much shorter than geological time scales and environmental change may result in considerable human pain and suffering. We are deeply attached to our home environment and need to see it as it is, so then appropriate changes can be made. Perhaps this emotional element coupled with objective scientific thinking will result in deciding the most important questions to ask.

In Steven Johnson's book, *The Invention of Air,* he states "Countless other cultures have imagined themselves living at the apex of history and human understanding. Priestley took that assumption, grounded it into an empirical story of scientific discovery, and then added the crucial caveat: 'This is only the beginning!'"[4]

Great solutions begin with great dreams and what follows are dreams of community, scientific and technological realities yet to exist coupled with a new human awareness for a truly sustainable future. Over the last 500 years there has been a new world-view created where the earth-centered solar system has been replaced with one where the sun is in the center. This is despite our senses seeing the sun rising in the east and setting in the west. Now, in the development of human consciousness, we need to move away from the human-centered mind set to a more inclusive ecological consciousness where we are part of a greater whole.

Many environmentalists, among them Bill McKibben, scholar-in-residence at Middlebury College, support the idea that the future is about consuming less and making more community. In his book, *Deep Economy,* McKibben provides the example of eating at a local restaurant that buys food from the organic farmer down the road. By eating at the local restaurant you support both the local business and the local farmer. In essence you are supporting your community and living within your own bioregion. We also live in a "flat world" described by Thomas Friedman, New York Times columnist, where many of us are working on our laptops with people half way around the world; but that does not have to mean we don't know our neighbors! Consider a future that is not just about the *more* stuff we have the *better* we feel. Our earth cannot sustain the More and Better (from McKibben's *Deep Economy*) mindset. The economy of future also needs to be about the quality of life.

To add a sense of balance in the context of "More" and "Better," McKibben began his recent book with the following story of a Chinese girl:

> *"Recently I was on a reporting trip to China, where I met a twelve-year-old girl named Zhao Lin Tao, who was the same age as my daughter and who lived in a poor rural village in Sichuan province – that is, she's about the most statistically average person on earth. Zhao was the one person in her crowded village I could talk to without an interpreter: she was proudly speaking the pretty good English she'd learned in the overcrowded village school. When I asked her about her life, though, she was soon in tears: her mother had gone to the city to work in a factory and never returned abandoning her and her sister to their father, who beat them regularly because they were not boys. Because Zhao's mother was away the authorities were taking care of her school fees until ninth grade, but after that there would be no money to pay. Her sister had already given up and dropped out. In Zhao's world, in other words, it's perfectly plausible that More and Better still share a nest. Any solution we consider has to contain some answer for her tears. Her story hovers over this whole enterprise. She's a potent reality check.*

> *And in the end it's reality I want to deal with – the reality of what our world can provide, the reality of what we actually want. The old realism – an endless More – is morphing into a dangerous fantasy.... In the face of energy shortage, of global warming, and of the vague but growing sense that we are not as alive and connected as we want to be, I think we've started to grope for what might come next. And just in time."*[4]

Thomas Berry, philosopher, environmentalist, theologian, and writer (1914-2009) has articulated the broad challenges that need to be scaled in order for us to come back into a state of balance with the natural world in *The Great Work.*

Consider these following three ideas of Thomas Berry as specific next steps:

> First, we need to implement the idea of bioregionalism. A bioregion is the place where a region's water comes from. Think of geographic regions by their particular ecology. Wherever around the world you pick, (Asia, Middle East, Africa, America) the only sustainable solution is a bioregional one.
>
> Second, let us rekindle our respect for the earth and acknowledge that the earth is primary and the human is derivative. One way we can demonstrate this is to reassess the cost of "stuff." We have been undervaluing the environmental costs of goods and our planet has suffered.
>
> Third, there will be a need for epic developments in ritual, art, drama, and literature. These are the human activities that feed the soul and are even more critical in times of intense change.

Lastly, here are some technological ideas/dreams. Consider the price of solar energy dropping and with technology innovation all of the world's villages have access to solar energy. Now couple this with the Internet providing opportunities for people around the world to share information. They can also sell their goods, thus improving their quality of life. Implementation of these ideas may stem the migration of millions of people from rural villages into the squalor of megacities and improve countless lives. Genetic engineering of plants can improve the efficiency of photosynthesis from 1% to 10%, thus increasing growth of the trees and reducing atmospheric carbon dioxide. Fully understanding how a seed develops into an organism and how these organisms interact with the ecology can be studied in depth so we may also be able to recreate entire ecosystems.

"It is likely that biotechnology will dominate our lives and our economic activities during the second half of the twenty-first century, just as computer technology dominated our lives and our economy during the second half of the twentieth. Biotechnology could be the great equalizer spreading wealth over the world wherever there is land and air and water and sunlight... In a world economy based on biotechnology some low cost and environmentally benign backstop to carbon emissions is likely to become a reality. ... Environmentalism, as a religion of hope and respect for nature, is here to stay. This is a religion that we can all share, whether or not we believe that global warming is harmful. ...Many of the people who are skeptical about the harmfulness of global warming are passionate environmentalists. They are horrified to see the obsession with global warming distracting public attention from what they see as more serious and more immediate dangers to the planet, including problems of nuclear weaponry, environmental degradation, and social injustice. Whether they turn out to be right or wrong, their arguments on these issues deserve to be heard.[5]

~Freeman Dyson

But we must always be open to the surprises of science. It often moves forward NOT by careful planning. This is the way invention works. In the real world nothing is predictable and success is usually a surprise.[6]

Whatever the project, consider the following seven steps:

First, identify your problem and study it in meticulous detail.

Second, develop hypotheses for the successful implementation of your plan.

Third, plan for contingencies if your plan should fail.

Fourth, perform a pilot to see if on a small scale your predictions come true in practice.

Fifth, continue to refine your plan as it becomes implemented still on a small scale.

Sixth, educate the public as to your thinking and the key points that will result in success.

Seventh, continue to implement your plan with growing buy-in and support as the early subscribers to your thinking are shown to benefit.

You have learned the thought processes in the scientists' minds. Now, consider the potential synergy when integrating the work of scientists, philosophers/theologians, activists, and engineers. When we think of the world, rather than thinking solely in terms of national boundaries we should try to think bio-regionally. We will need to find both solutions to scientific and technological challenges and then solutions at the level of implementation. In both cases the scientific method is a path forward. This is a slow, patient, and deliberate process. Integrate this thinking with the New Story of Thomas Berry where we have "a comprehensive ethics of reverence for all life." Leave your bias at the door and I believe the wings of creativity and the potential of the human mind are up to the task.

Let's get to work!

Reflection

The history of science is full of stories of innovation born of rejecting the status quo; discoveries don't get made if scientists don't look for things to discover. From Leavitt's Cepheids, to Rutherford's atom, to planetary theory, to Darwin's evolution, it is clear that scientific discovery is most advanced when scientists look for new ways to interpret ideas. In this context, there is no reason for us as a nation to not approach the environmental crisis with the same open-mindedness as demonstrated in all of the disciplines of science.

AN INTERVIEW WITH FREEMAN DYSON

As discussed throughout this book, science is both a fascinating collection of facts and a method for inquiring about the natural world. To end this book I have comments by one of the brightest scientific minds of the twentieth century. In November 2011, Freeman Dyson attended my class titled, *The Universe and the Origin of Life*, at the Westminster Choir College of Rider University. Professor Dyson is an Emeritus Professor of mathematics and theoretical physics at the Institute for Advanced Study in Princeton, New Jersey. He is best known for his work in the area of physics called quantum electrodynamics. In addition to his work in science and mathematics, Professor Dyson has written books for the non-scientist. Titles of some of his books include *Infinite in All Directions*, *The Sun, the Genome and the Internet*, *The Scientist as Rebel*, and *Disturbing the Universe*. He is the recipient of many scientific awards, among them the Templeton Prize for his work on science and religion.

At its core, science is a curiosity-driven activity whose goals are to uncover the wonders of the natural world. Freeman's intelligence, humility, honesty, and sense of humor remind us of this fact.

Questions from Students to Freeman Dyson

Question/comment: I'm disappointed in the lack of manned space program because it is the loss of a frontier mindset. You can only get so much from collecting data from something. We haven't really learned about space and Mars until we have been there.

Dyson: As a human adventure the space program has been a sad story, but the science is going on fantastically well. Don't confuse the two. I visited the Russian launch center and my daughter was a back up cosmonaut to Charles Simonyi. She didn't get to go up into space (yet) but I saw the Russian way of doing space. Their culture is all together different from ours. Their view of space is that it is a human destiny and it will take centuries. In Baikanur, Kazakhstan, on the morning of a launch the whole town came out. Even though the day of the launch was rainy and windy, when the cosmonauts came out from quarantine they were paraded through the city to shouts and screams. The mayor made a long speech and the cosmonauts would say, "We are ready to fly." Then they would get in the rocket and go. All of this was done with great ceremony and not in a hurry. The Soyuz launcher is still used after fifty years. Why change it? They believe they are on the way to the stars. We need to change our way of thinking and Congress has to think longer than ten years ahead.

Question/comment: In the United States we have a lack of drive to look forward more than a very few years.

Dyson: The environmental movement looks forward more than a few years and this is movement in the right direction.

Question/comment: Do you like stargazing?

Dyson: Yes. Now that I'm getting old I feel the cold nights more. We have kids in California and they are high up in the mountains. The sky up there is unbelievable and it takes your breath away. In Princeton, you can hardly see the sky.

There is a project called telescopes in education, and a computer driven telescope can be controlled remotely by a classroom of kids. I had the joy of being in the dome of a telescope in California that was being controlled by kids in Virginia and you could hear their voices. Jupiter was under bombardment about twelve years ago and as a comet disintegrated its pieces fell into Jupiter. Huge blotches on the surface of Jupiter were created and then you could see Jupiter rotating. It was amazing watching Jupiter spin. Kids were shouting and grabbed hold of the telescope.

Question/comment: The Nobel Prize for Physics in 2011 was awarded for learning that the universe's expansion is accelerating. What is the universe expanding into?

Dyson: The simple answer is that we don't know. What we call the universe consists of what we can see and that is what we can know. What makes science interesting is that the world is full of mysteries full of what we don't know.

This was a very well earned prize. They were observing type Ia supernova and with improvements in telescopes we see hundreds of these supernovas every year. White dwarf stars blow up in a standard fashion with the same brightness each time. We can compare the intrinsic brightness with how bright the supernova looks and can calculate the distance to the object. Then, independently, we measure the velocity of its motion away from us using the spectral lines of the chemical elements in the star. From both the distance and velocity we can determine if the universal expansion is accelerating or decelerating. Because of the large number of supernovas that we can now observe we can know these rates of acceleration rather precisely. It is amazing how far away we can see objects, and the universe as a whole is pretty much transparent no matter how far away the object is. Sometimes things pile up and hide each other but not very often.

Question/comment: Experiments in Europe have recently suggested that the speed of neutrinos can be faster than the speed of light. What do you think about this?

Dyson: It is a lousy experiment. Here's why, using an analogy to long distance runners. For long distance runners there are two kinds of time, gun time and chip time. (On the sneakers there is a chip that is turned on when the runner crosses the start line and is turned off when the runner crosses the finish line. If there are many runners piled up so they don't all cross the start line at the same time then the gun time will be different from the chip time.) It is really the chip time that counts, but in the world of running committees make up the rules and the gun time is what is used rather than the chip time. In the neutrino experiment what was measured was the gun time and neutrinos don't carry chips. When the investigators turned on their machine they didn't know how far behind the "start line'" was the neutrinos. People at CERN are angry about this and the results were published by a press conference rather than by peer review. Public media has a tendency to either cure cancer or prove Einstein wrong. This needs to be discounted.

Question/comment: Most of history sees the universe as static, not dynamic. What do you think about that?

Dyson: Herschel surveyed the sky and said that he was looking back into time—not just far away. He said the universe was a chronometer that changed with time. This was about 1770 and it didn't get through to the public for another hundred years. Close to the beginning of the twentieth century was Einstein and relativity, and it was very much publicized that time is relative. Public knowledge of physics and astronomy was growing, and the public gave credit to Hubble but the idea that when we look at the sky we are looking back in time was really originally conceived by Herschel.

Question/comment: We have had some interesting conversations about the life of Rutherford and would you share the story with my students.

Dyson: Rutherford was an old-fashioned paternalistic professor. He took care of the lives of his students. He had strong views about how they should live their lives. They were not allowed in the lab after 6 p.m. and no one could come in to work for the rest of the day since that was the time for the family. They had to be human for the rest of the day. Also, he required them to take four serious two-week holidays per year and they were supposed to go off on holiday. This worked amazingly well. They did a tremendous amount of first-rate science, as much as Americans working past midnight and never taking holidays.

There was a race going to discover the nature of the atom between Rutherford and two other groups, Van der Graaff in Boston and Lawrence in Berkeley, California. They were building machines frantically to split the nucleus and Lawrence was famous as a slave driver (the opposite of Rutherford). The beauty of it is that Rutherford won the race. Don't work too hard.

Question/comment: Do you see any value to finding out how the origin of life happened?

Dyson: Oh yes. Science is all about connections. We don't know where it will lead to and will discover quite unexpected information that we could never plan. We are exploring back and back and back to where life originated, and along the way there will be connections to all sorts of other things.

Question/comment: Do you think the way we look at ourselves will change when we learn about the origin of life and explore space and life elsewhere in the universe?

Dyson: How we apply our knowledge is different from how we obtain our knowledge. So understanding the origin of life will be a huge accomplishment. As far as human nature is concerned, over a time scale of centuries, if we went to another planet then we might experiment with and change human nature but as long as we stay here on earth we may as well stay as we are.

Question/comment: You wrote Origins of Life in which you speculate about two origins, the origin of metabolism and the origin of replication. What has been found out in the last twenty-five years?

Dyson: We're still very far away from understanding the origin of life. Most primitive creatures are still vastly more complicated than what must have happened at the origin. There is a dark age that we haven't been able to penetrate between the origin of life and what is present now. No progress on that. Doron Lancet at the Weitzman Institute does careful computer modeling on the origin of life. He has a lot of chemistry in his model, but computer models are no measure of reality. To make real progress we need to work from the test tube and simple chemical experiments. It is very hard. Take a million globules and put "garbage" into them and see what happens. The technology to do that is becoming available, and we are probably 100 years away from understanding how it works.

Question/comment: What about the intersection of science and the humanities?

Dyson: There has been a big penetration by science into our view of ourselves. Galileo did for astronomy what Darwin did for biology. Darwin did more to influence our view of ourselves than astronomy.

Question/comment: Expand on your thoughts about human nature.

Dyson: I was thinking a bit recently about evolution of our minds because of reading *Thinking Fast and Slow*, by Daniel Kahneman, Nobel Laureate in economics. Kahneman is a psychologist who thinks about economics. He has a notion of two parts of our brains: the quick and the slow. Both are essential and are quite separate. Quick brain deals with decisions we have to make instantaneously. We recognize faces in a fraction of a second and we hear a noise and know what it is. We know in a fraction of a second if there is a problem. The quick brain makes mistakes and we often get things wrong. The second half of the brain is slow and corrects and analyzes mistakes. This is also very important but most of the time we use the fast brain. How did these two parts of the brain evolve? One hundred million years ago the fast brain was evolving in mice and we were in danger of being eaten by reptiles. We used our brain to be aware of danger quickly enough to escape and get our kids into the hole. This is how the fast brain evolved in dense jungle. Five million years ago we became apes and lived in trees. Now we didn't have to worry about predators all of the time. We worry about territory and fruit growing on a tree that needs to be part of our territory. So life is essentially a matter of territory and this does not have to be fast. The brain can be slow and discuss what to do. This is a poetic picture. The mouse brain is the fast brain and the ape brain is the slow brain. Two million years ago we came down from the trees and since that time we have been getting along with both of them together. The environment really does have a large effect on what we are. Hole in the ground (mouse) and top of the tree (ape) and we're in the middle.

Question/comment: Everything is the same – connections are what make knowledge useful. In the end everything boils down to a structure and function. In the cell most of the time if something different happens then the cell dies. But sometimes when something different happens evolution occurs. What about at the level of anthropology and culture?

Dyson: My friend Martin Novak studied the evolution of cooperation. He studied how do you get a society to cooperate when it is advantageous to the individual to cheat. How did cooperation evolve? Obviously the person who cheats will evolve and have more kids. Punishment is the way of controlling cheaters and bullies will punish the cheaters. What goes with that and is most illuminating is that we enjoy punishing the cheaters. There was an experiment where people had to pay to punish cheaters. Artificial situations were set up and people paid as much as 1/3 of their winnings to punish cheaters.

Question/comment: Do you think there is a relationship between science and music?

Dyson: They are both about learning how to use a set of tools. The beauty of both is they are marvelous tools and it is amazing what they do if you learn how to use them. Science is fundamentally different since science is a much more public enterprise whereas music is private. My father wrote big choral works and that is more like science where big groups of people perform together.

Thank you for asking good questions.

TIMELINE FROM THE BIG BANG TO THE ORIGIN OF CULTURE

Big Bang (13.7 BYA)

13.7 billion years ago there was a Big Bang that resulted in the creation of all of the matter in the universe. This was the beginning of both time and space.

Creation of Matter 13.7 BYA)

The temperature was so hot --literally billions of degrees -- that the nature of the matter was unlike anything that we know on earth today.

Creation of Galaxies (13.3BYA)

The universe continued to expand, but there were local regions of density that ultimately resulted in the creation of galaxies. The force of gravity causes the local regions of density to continue to aggregate together and ultimately stars were formed.

Creation of Chemical Elements (10BYA to present)

If a star is about ten times the mass of our sun, then the rate at which hydrogen fuses into helium happens faster than in our sun and the star runs out of fuel in millions rather than billions of years. As the star continues to contract, the energy results in the fusion of helium and all of the atoms of the periodic table are produced. At the time of maximum compression the conditions are right for an explosion of the elements into space and we have a supernova explosion. One such supernova exploded in the region of the Milky Way galaxy out of which our star and solar system were formed about 4.5 billion years ago.

Creation of our Sun, Earth, and Solar System (4.5 BYA)

Ninety-nine percent of the matter that created the sun and solar system is hydrogen and helium and it made up the sun. The remaining 1% formed the inner rocky planets (Mercury, Venus, Earth, and Mars) and the outer gaseous planets (Jupiter, Saturn, Uranus, and Neptune). The distance of the earth from the sun as well as the spinning of the earth on its axis enabled conditions to permit the evolution of life. The atmosphere contained no oxygen and may have been created by volcanic explosions. The center of the earth, consisting of spinning liquid iron, has a temperature of thousands of degrees and creates a magnetic field around the earth. These are the geological properties of a living planet.

Origin of Biomolecules and Life (3.5 BYA)

The origin of life may have occurred in thermal vents under the ocean that derive their energy from the heat of the earth. Alternatively it occurred in a pond on the surface of the earth. The environmental conditions of the early earth may have been thousands of times more extreme than our most intense thunder and lightning storms. All we know is that it happened!

Transformation of the Earth's Atmosphere (2.5 BYA)

About 2.5 billion years ago there was the origin of photosynthesis where plant cells were able to take carbon dioxide from the atmosphere and convert it into sugar necessary for the energy of life. A byproduct of photosynthesis is oxygen -- the source of all atmospheric oxygen today. At the top of our atmosphere, the ultraviolet light from the sun reacts with oxygen creating the ozone layer that blocks out the most dangerous ultraviolet rays. This resulted in a less extreme environment in which subsequent evolution occurred.

Evolution of Bacteria, Plants, and Animals (2 BYA to present)

In the presence of oxygen many cells died because of oxygen toxicity, but those that survived were able to use the oxygen effectively to create more energetic forms of life. The similarity of biochemistry between bacteria, plants, and animals fills thousands of pages of textbooks and is further proof of the evolutionary relationship between all forms of life.

With the evolution of multicellular creatures came a bounty of new life forms referred to as the Cambrian explosion. Many different forms of fish as well as land plants and animals evolved at this time.

Evolution of the Dinosaurs (250 MYA)

Among the adaptations about 250 million years ago were mutations that resulted in the evolution of dinosaurs from lizards. They lived on the earth for about 200 million years and the small dinosaurs evolved into birds.

Mass Extinction (65 MYA)

About sixty-five million years ago there was a meteorite that hit the earth in the vicinity of the Yucatan peninsula. This meteorite was about 10 km in diameter and hit the earth with such force that there was a global catastrophe resulting in the extinction of many forms of life including the dinosaurs. Among the groups of animals that survived were the small mammals from which we evolved.

Evolution and Migration of Humans (3 MYA)

About three million years ago there were about thirty different species of hominids from which humans evolved. Only one species, Homo sapiens, survived and we are all descended from this species with a small admixture of Neanderthals and other species. It is thought that our predecessors originated in Africa and then they migrated into Asia and Europe. Finally they came over the Bering land bridge to Alaska and then down through North and South America.

Origin of Culture (10,000 years ago)

Finally we have the origin of culture about 10,000 years ago and the creation of written language. We have finally been able to evolve to a point where we can ponder our own existence and study the nature of our origins. After all of this time we become aware that we are made of the stuff of stars and are intimately related to our environment and all other living things on the earth.

QUESTIONS TO CONSIDER

Some questions are more philosophical and others are more technical. Complete answers may require additional research beyond the text (see Bibliography). For technical questions consider both WHAT we know and HOW we know. When you provide a scientific answer to a question include a testable hypothesis in your answer. Also, think about the following:

1. Is the evidence due to *causation* or is the evidence a *correlation*? A way to answer this is to pose an if/then statement to determine if the answer is always true or not. Proving causality is not easy.

2. Is the answer based on observations or based on inference from observations? If the tools that one is using are good then we can assume the observations are correct. Inferences from observations are more subject to an individual point of view.

Name ______________________________ Date ______________

Introduction

1. Give an example of how data-based decision making and constructive criticism benefits our culture.

2. Where do you think human beings come from?

3. What do you aspire for in your education? What is the role of reading, writing, mathematics, and critical thinking?

4. Do you think science has affected religion? How?

5. Review a statement that you have heard lately on TV, the radio, or the internet. How do you know if the statement is true?

6. Do you agree with Sister Miriam MacGillis' story that we need an alternate story to the one where we can extract all we want from the earth with no bad consequences? Explain.

7. What does it mean to have a "comprehensive ethics and reverence for all life"?

8. Paraphrase what is meant by humanism. Is your opinion about the value of science affected by your opinion of humanism?

9. Do you believe humans are part of nature?

10. Comment on the following reflection and what thoughts does it trigger in your own mind: *All it takes is a little knowledge, scientific thought, and open-mindedness. If there is one common thread among the vastly diverse disciplines of science, it is about being open-minded and questioning as much as it is about new discoveries.*

Name ____________________ Date ____________

Astronomy

1. Compare the Aristotelian and the Copernican views of the universe. Give the dates for how long the Aristotelian view lasted and when the Copernican view began.

2. If science changes over time, then why can we believe a scientific thought more than a random guess or whim?

3. Why were the observations of Tycho Brahe of a supernova explosion and a comet so damaging to the Aristotelian theory of the universe?

4. Review the logic and do a calculation for how we know that the universe is so large.

5. How does analysis of the spectral lines from galaxies tell us that we live in an expanding universe?

6. What was the metaphor that Martin Rees used to describe the size of our solar system and the distance to the nearest star?
(PAGE 11)

7. Review what we think were the events in the formation of our solar system beginning with a supernova exploding in the neighborhood where our solar system is located.

8. What is the relationship between the size of a star and its lifetime? Why is our sun more suitable to a solar system in which life evolved compared to a star that is ten times larger?

9. Comment on the following reflection, give an example in your own life, and what thoughts does it trigger in your own mind? "*dare to be hungry for knowing, and better ourselves through understanding.*"

Name ____________________ Date ____________

Chemistry

1. Give several examples of how developments in chemistry impacted civilization.

2. Review the experiment of Priestley where he showed the relationship between plants and animals and combustion.

3. Review the experiment where Lavoisier showed that air contained ~20% oxygen.

4. Review the experiment of Rutherford in which he concluded that the structure of an atom had a lot of empty space and a tiny dense nucleus.

5. Review the evidence that the origin of the elements is inside stars. What does this tell us about how stars shine?

6. Why is the Miller and Urey experiment important in linking the study of chemistry to the study of biology?

7. Retell the story of how a carbon atom cycles through all of life by the processes of photosynthesis and respiration. Where do coal, oil, and natural gas come from?

8. Comment on the following reflection and think of an idea that questions the status quo. *It was those scientists who questioned the status quo that came forth with new innovations*

Name ______________________________ Date ______________

Geology

1. Review the steps for the formation of the earth. What is the function of liquid metallic iron and nickel in the core?

2. What evidence is there that the earth is very dynamic and powerful (even if you do not experience an earthquake or witness a volcanic explosion)?

3. Review the evidence that the continents were once a supercontinent (Pangaea). Where do geologists think the energy comes from that drives the movement of the continents?

4. How are the seismic waves, triggered by earthquakes, used to discover the structure of the earth?

5. What is the evidence and significance that meteorites have impacted the earth?

6. What is the effect of the moon on the earth and where do we believe the moon came from?

7. Review the thinking of Gould where he states that if we rewind the clock and evolve life all over again we would almost certainly evolve different species – and not us.

8. Comment on the reflection --*though life has changed the environment, we have the environment to thank for life.*

Name ______________________ Date ______________

Biology

1. Why do we think it was important that Darwin had read Lyell as well as Malthus in formulating his theory of evolution by natural selection?

2. Define metabolism and replication and how they are important in the origin of life.

3. How do we know that mutation, variation, and natural selection drive evolution?

4. Discuss the importance of the evolution of photosynthesis for the history of life on earth and its environment.

5. Discuss the importance of the evolution of sexual reproduction for the history of life on earth.

6. How do we know that we evolved from small mammals that lived about eighty million years ago?

7. Do you agree with Gleiser when he says the existence of humans is more than a pointless accident in an indifferent universe?

8. Comment on the following reflection and what thoughts does it trigger in your own mind. *Even though it took nine months to create you, it took millions and millions of years of mutations and evolution to create humans.*

Name ____ Date _______________________

Sustainability

1. What thoughts come to mind when you read the quote: "…our home must be defended like a holy relic"?

2. Have you ever experienced *solostalgia*?

3. Why does it take so long for new ideas to be accepted into society?

4. What do you think are the causes of the increasing carbon dioxide in the atmosphere over the last fifty years?

5. What might be some solutions to our environmental problems?

6. What does the word biotechnology mean?

7. What does McKibben mean by "More" and "Better" being replaced by more community and connectedness among people?

8. What does Thomas Berry mean by bioregionalism?

9. Comment on the reflection, do you agree with it, and what thoughts does it trigger in your own mind? *Does it matter that we know the Story of the Universe? The story allows us to appreciate the uniqueness of the earth. This appreciation leads to study and research. Research leads to data and understanding. Understanding leads to predictions. Predictions can lead to present-day action.*

Name ______________________________ Date ______________

Dyson Interview

1. How does the Russian way of supporting their space program contribute to its long-term sustainability?

2. What is the value of understanding how life originated?

3. What does Dyson say about human nature and the importance of the fast and slow parts of the brain?

4. How do you get a society to cooperate when it is advantageous for an individual to cheat?

BIBLIOGRAPHY

Astronomy

Books

Forbes, George. *History of Astronomy*, Putnam, New York, 1909.

Hubble, Edwin. *The Observational Approach to Cosmology*, Oxford at the Clarendon Press, 1937.

Rees, Martin. *Just Six Numbers, The Deep Forces that Shape the Universe*. Weidenfeld and Nicolson, London, 1999.

Sagan, Carl, *Cosmos*, Random House, New York, 1980.

Wilson, Robert. *Astronomy through the Ages: The story of the human attempt to understand the Universe*. Princeton University Press, 1997.

Articles

Burbidge, E.M., Burbidge, G.R., Fowler, W.A. and Hoyle, F. Synthesis of the Elements in Stars. *Reviews of Modern Physics*, Volume 29, number 4, October 1957, pp 548-647.

Web sites and videos

"Electromagnetic Radiation (Light)." http://www.astronomynotes.com/light/s1.htm. Description of light as an electromagnetic wave

"Powers of Ten." 1997. http://www.powersof10.com/film. Film depicting the size of the universe from the very large to the very small

"Red Shift." http://hyperphysics.phy-astr.gsu.edu/hbase/astro/redshf.html#c1. Physics underlying the red-shift

Sagan, Carl. "The Pale Blue Dot." http://www.edutube.org/en/video/carl-sagan-pale-blue-dot-full-speech. Speech of Astronomer Carl Sagan

"The Wilkinson Microwave Anisotropy Probe (WMAP)." http://hyperphysics.phy-astr.gsu.edu/hbase/astro/wmap.html#c1. Discussion of the microwave background radiation

Chemistry

Books

Isaac Asimov. *Atom: Journey across the Subatomic Cosmos*, Truman Talley Books, Dutton, New York. 1992.

Brock, William H. *The Norton History of Chemistry*. New York and London: WW. Norton and Company, Ind., 1992.

Kean, Sam. *The Disappearing Spoon: And other True Tales of Madness, Love, and the History of the World from the Periodic Table of the Elements*. New York: Little, Brown and Company, 2010.

Levey, Martin. *Chemistry and Chemical Technology in Ancient Mesopotamia*. Amsterdam, London, New York, Princeton: Elsevier Publishing Company, 1959.

Levi, Primo. *The Periodic Table* (English Translation). New York: Random House, 1984.

Read, John. *Through Alchemy to Chemistry*. London: G. Bell and Sons, Ltd., 1957.

Articles

Burbidge, E.M. J., G.R. Burbidge, W.A. Fowler, and F. Hoyle. "Synthesis of the Elements in Stars." *Reviews of Modern Physics* 9, no. 4 (October 1957): 548-647

Web sites

"Miller/Urey Experiment." http://www.chem.duke.edu/~jds/cruise_chem/Exobiology/miller.html. Information on the Miller and Urey Experiment

"Rutherford's Experiment: Nuclear Atom." http://www.youtube.com/watch?v=5pZj0u_XMbc. Video of the Rutherford Experiment

Geology

Books

Faure, G. *Principles and Applications of Geochemistry, 2nd Edition*. New Jersey: Prentice Hall, 1998.

Gould, Stephen J. *Wonderful Life*. New York: W.W. Norton and Company, 1989.

Lyell, Charles. *Principles of Geology* Vol 1, London, John Murray, 1830.

Stradins, Ina, Angeles G. Guerrero, and Frances Peter, eds. *Prehistoric Life*. London: Dorling Kindersley, 2009.

Yan, Hong-Sen. *Reconstruction designs of lost ancient Chinese machinery*. Dordrecht : Springer, 2007.

Articles

Miller, K.G. et al. "Relationship between mass extinction and iridium across the Cretaceous-Paleogene boundary in New Jersey." *Geology* 39 (October 2010): 867-80.

Web sites

"Exploring the Earth Using Seismology." http://www.iris.edu/hq/resource/exploring_the_earth_using_seismology More information on the structure of the earth and the use of seismology

"Seafloor Spreading." http://platetectonics.pwnet.org/story_tectonics/theory/seafloor_spreading.htm. More information on sea floor spreading

inventors.about.com/library/inventors/blseismograph2.htm Zhang Heng's seismoscope.

Biology

Books

Alberts, Bruce, Alexander Johnson, Julian Lewis, Martin Raff, Keith Roberts, and Peter Walter. *Molecular Biology of the Cell, 4th Edition*. New York: Garland, 2002.

Darwin, Charles. *The Origin of Species by Means of Natural Selection or The Preservation of Favoured Races in the Struggle for Life*. London: J. Murray, 1859.

Dawkins, Richard. *The Blind Watchmaker: Why the evidence for evolution reveals a Universe Without Design*. New York: W.W. Norton and Company, 1986.

Dawkins, Richard. *The Greatest Show on Earth: The Evidence for Evolution*. London: Free Press, 2009.

Dyson, Freeman, *Origins of Life*. Cambridge: Cambridge University Press, 1985.

Mattock, J.N. and M.C. Lyons. (edited and translated from Arabic). *Hippocrates: On Endemic Diseases (Airs, Waters and Places)*, Volume 5. Cambridge: W. Heffer and Sons Ltd., 1969.

Meinesz, Alexandre (translated by Daniel Simberloff). *How Life Began: Evolution's Three Geneses*. Chicago: The University of Chicago Press, 2008.

Schopf, J. William. *The Discovery of Earth's Earliest Fossils*. Princeton: Princeton University Press, 1999.

Schopf, William J., ed. *Life's Origin: The Beginnings of Biological Evolution*. Los Angeles: University of California Press, 2002.

White, M.J.D. *The Chromosomes*. New York: John Wiley and Sons, Inc., 1937.

Articles

Petroli, R.J., A.T. Maciel-Guerra, F.C. Soardi, F.L. de Calais, G. Guerra-Junior, and M.P. de Mello. "Severe forms of partial androgen insensitivity syndrome due to p.L830F novel mutation in androgen receptor gene in a Brazilian family." *BMC Research Notes* 4, no. 173 (2011).

Web sites

"Darwin Online." http://darwin-online.org.uk/graphics/Origin_Illustrations.html. Charles Darwin reference materials

"How did life on earth begin?." http://www.youtube.com/watch?NR=1&feature=endscreen&v=dYphAH2tKYE. Tree of life video

Szostak, Jack. "The Origin of Life - Abiogenesis." http://www.youtube.com/watch?v=U6QYDdgP9eg. Origin of life video by Jack Szostak

http://www.nap.edu/catalog.php?record_id=11876 Science, Evolution and Creationism

Sustainability

Books

Berry, Thomas. *Dream of the Earth*. San Francisco: Sierra Club Books, 1988.

Berry, Thomas. *The Great Work: Our Way into the Future*. New York: Bell Tower, 1999.

Brown, Lester. *Plan B 4.0: Mobilizing to Save Civilization.* London: W. W. Norton and Co., 2009.

Brown, Lester. *World on the Edge: How to Prevent Environmental and Economic Collapse.* New York: W.W. Norton and Company, 2011.

Crist, E. and Bruce H. Rinker, eds. *Gaia in Turmoil: Climate Change, Biodepletion, and Earth Ethics in an Age of Crisis.* Boston: MIT Press, 2010.

Dyson, Freeman. *The Sun, The Genome and The Internet.* New York: Oxford University Press, 1999.

Leopold, Aldo. *A Sand County Almanac.* New York: Oxford University Press, 1966.

McKibben, Bill. *Deep Economy: The Wealth of Communities and the Durable Future.* New York: Henry Holt and Company, 2007.

McKibben, Bill. *Eaarth: Making Life on a Tough New Planet.* New York: Holt and Company, 2010.

Murphy, Pat. Plan C: *Community Survival Strategies for Peak Oil and Climate* Change. Gabriola Island, BC, Canada: New Society Publishers, 2008.

Swimme, Brian and Thomas Berry. *The Universe Story.* San Francisco: HarperCollins, 1992.

Article

Albrecht, Glenn. "Solastalgia: the distress caused by environmental change." *Australiasian Psychiatry* 15 (2007): S95-98.

Web sites

"CO_2 Concentration at Mauna Loa Observatory, Hawaii." http://scrippsco2.ucsd.edu/. Keeling Curve of carbon dioxide increasing in the atmosphere over the last fifty years

"Thomas Berry." http://www.thomasberry.org/Biography/tucker-bio.html Biography of Thomas Berry

"The Question of Global Warming." http://www.nybooks.com/articles/archives/2008/jun/12/the-question-of-global-warming/?pagination=false. Freeman Dyson article in the New York Review of Books on the global warming

"Wedges Reaffirmed." http://www.climatecentral.org/blogs/wedges-reaffirmed/. A discussion on ways to stabilize fossil fuel emissions

General science

Angier, Natalie. *The Canon: A Whirligig Tour of the Beautiful Basics of Science.* New York: Houghton Mifflin Company, 2007.

Bruno, Leonard C. *The Tradition of Science, Landmarks of Western Science in the Collections of the Library of Congress.* Washington, DC: Library of Congress, 1987.

Bryson, Bill. *A Short History of Nearly Everything.* New York: Broadway Books, 2003.

Dyson, Freeman, *The Scientist as Rebel.* New York: New York Review of Books, 2006.

Ferngren, Gary, Edward Larson, Darrell Amuindsen and Anne-Marie Nakhla, eds. *The History of Science and Religion in the Western Tradition.* New York and London: Garland Publishing, Inc., 2000.

Fine, Cordelia. *A Mind of its Own: How your Brain Distorts and Deceives.* New York: W. W. Norton and Company, 2006.

Johnson, Steven. *The Invention of Air.* New York: Riverhead Books, Penguin Group (USA), 2008.

Judson, Horace Freeland. *The Search for Solutions.* New York: Holt, Rinehart and Winston, 1980.

Lodge, Sir Oliver. *Pioneers of Science.* London: Macmillan and Co. Limited, 1905.

Other books

Eco Umberto, translated by William Weaver, *Serendipities: Language and Lunacy.* New York: Harcourt, Brace and Company, 1999,

Huxley, Aldous. *Island.* New York: Harper and Row, 1962.

Tzu, Lao, (translated by Gia Fu Feng and Jane English). *The Tao Te Ching.* New York: Random House, 1972.

LIST OF FIGURES WITH CITATIONS

Figure 3.1 Turtle shell fossil
(Owen, Richard. *History of British Fossil Reptiles.* London: Cassell, 1849)

Figure 3.2 Wegener's diagram of continental movement
(From Bruno, p. 220)

Figure 3.3 Earth with quake waves emanating from Japan. Image © jpeg-wiz, 2013. Used under license from Shutterstock, Inc.

Figure 3.4 Seismic Diagram illustrating seismic waves arriving at different stations. Image © photopixel, 2013. Used under license from Shutterstock, Inc.

Figure 3.5 Seismic wave velocity graph
FAURE, GUNTER, PRINCIPLES AND APPLICATIONS OF GEOCHEMISTRY, 2nd Edition, © 1998. Reprinted by permission of Pearson Education, Inc., Upper Saddle River, NJ.

Figure 3.6 Craters on the moon
(Lodge, p. 97)

Figure 4.1 Gibbon with baby
St Hilaire, I.G. Description des mammiferes nouveau; ou,imparfaitement connus de la collection du museum d'histoire naturelle, et remarques sur la classification et les characteres des mammiferes. Paris: Gide, 1841-1861.

Figure 4.2 Man's place in nature
Huxley, Thomas Henry, *Man's Place in Nature*, 1863.

Figure 4.3 Androgen receptor amino acid sequences
From "Severe forms of partial androgen insensitivity syndrome due to p.L830F novel mutation in androgen receptor gene in a Brazilian family" by R.J. Petroli, A.T. Marciel-Guerra, F.C. Soardi, F.L. de Calias, G. Guerra-Junior, and M.P.Mello. BMC Research Notes, 2011, 4;173.

Figure 5.1 Devastated environmental picture. Image © Eugenio Marongiu, 2013. Used under license from Shutterstock, Inc.

Figure 5.2 Pictures of Earth from Space. Image © Sailorr, 2013. Used under license from Shutterstock, Inc.

Figure 5.3 Atmospheric carbon dioxide (1958-2008)
Source: Scripps CO_2 program.

FOOTNOTES

Introduction

1. Michael Cronon, *The American Scholar* (Volume 67, #4, 1998).
2. Freeman Dyson, in discussion with the author, November 2011.
3. Jared Flesher, "A Natural Calling," *Edible Jersey* (Spring 2011).
4. http://www.thomasberry.org/Biography/tucker-bio.html, accessed January, 2012.
5. Berry, 1988, p. 137.
6. Weldon, P. in Stephen Weldon in Gary Ferngren, Edward Larson, Darrell Amuindsen and Anne-Marie Nakhla, eds. *The History of Science and Religion in the Western Tradition* (New York and London: Garland Publishing, Inc., 2000), 212-213.

Chapter 1

1. Carl Sagan, *Cosmos* (New York: Random House, 1980), 4.
2. Wilson, *Astronomy Through the Ages*, 51.
3. Martin Rees, *Just Six Numbers, The Deep Forces that Shape the Universe* (London: Weidenfeld and Nicolson, 1999), 22.
4. Rees, *Six Numbers*, 82.

Chapter 2

1. E.M.J. Burbidge, G.R. Burbidge, W.A. Fowler, and F. Hoyle, "Synthesis of the Elements in Stars," *Reviews of Modern Physics* 9, no. 4 (October 1957), p. 548.
2. Sam Kean, *The Disappearing Spoon: And other True Tales of Madness, Love, and the History of the World from the Periodic Table of the Elements* (New York: Little, Brown and Company, 2010), 22.
3. John Read, *Through Alchemy to Chemistry* (London: G. Bell and Sons, Ltd., 1957), 142.
4. Asimov, p. 17
5. Primo Levi, The *Periodic Table* (English Translation), (New York: Random House, 1984), 224.

Chapter 3

1. Schopf, 1999. p. 167.
2. Yan, pp 129-130
3. Stephen J. Gould, *Wonderful Life* (New York: W.W. Norton and Company, 1989), 51.

Chapter 4

1. François Jacob, "The Birth of the Operon," Science 332 (May 13, 2011): 767.
2. J.N. Mattock, J.N. and M.C. Lyons, (edited and translated from Arabic), *Hippocrates: On Endemic Diseases (Airs, Waters and Places)*, Volume 5 (Cambridge: W. Heffer and Sons Ltd., 1969), 118-122.
3. Umberto Eco, p. 55.
4. Leonard C. Bruno, *The Tradition of Science, Landmarks of Western Science in the Collections of the Library of Congress* (Washington, DC: Library of Congress, 1987), 113.
5. Dyson, Origins of Life, 1985.
6. Richard Dawkins, *The Greatest Show on Earth: The Evidence for Evolution* (London: Free Press, 2009), 24-26.
7. Theodosius Dobzhansky, "Nothing in Biology Makes Sense Except in the Light of Evolution. *American Biology Teacher* (volume 35, 1973), p. 125.
8. D.W. Hillis, "Intelligence as an Emergent Behavior; or The Songs of Eden," *Dædalus, Journal of the American Academy of Arts and Sciences, special issue on Artificial Intelligence* (Winter 1988), 175-189.
9. Richard Dawkins, *The Blind Watchmaker, 1986, p. 159.*
10. Marcelo Gleiser, "We are Unique" in *This will make you Smarter, John Brockman, ed.,* (New York: HarperCollins, 2012), 5.

Chapter 5

1. Alice Calaprice, *The Quotable Einstein* (Princeton: Princeton University Press, 1996), 83.
2. Kevin W. Kelley, *The Home Planet* (New York: Addison Wesley Publishing Company, 1988), 24.
3. "Wedges Reaffirmed." http://www.climatecentral.org/blogs/wedges-reaffirmed/.

4. McKibben, *Deep Economy: The Wealth of Communities and the Durable Future* 2007, p. 4.
5. Dyson, 2008 The Question of Global Warming in the New York Review of Books, June 12, 2008
6. Freeman Dyson, in discussion with the author, May 2012.

INDEX

I

J

K

L

M

N

O